Also from Westphalia Press
westphaliapress.org

Our Engines of War, and How We Got to Make Them

Our Engines of War, and How We Got to Make Them

Westphalia Press
An imprint of Policy Studies Organization
1527 New Hampshire Ave., NW
Washington, D.C. 20036
info@ipsonet.org

ISBN-13: 978-1-63723-505-8

Cover design by Jeffrey Barnes:
jbarnesbook.design

Daniel Gutierrez-Sandoval, Executive Director
PSO and Westphalia Press

Updated material and comments on this edition
can be found at the Westphalia Press website:
www.westphaliapress.org

Our Engines of War, and How We Got to Make Them

by Capt. Jervis-White Jervis

WESTPHALIA PRESS
An imprint of
Policy Studies Organization

The Great Gun, Fifteenth Century. Leonardo da Vinci.

OUR

ENGINES OF WAR,

AND

HOW WE GOT TO MAKE THEM.

BY

CAPT. JERVIS-WHITE JERVIS, M.P.

ROYAL ARTILLERY,

AUTHOR OF THE "MANUAL OF FIELD OPERATIONS," "THE RIFLE-MUSKET," ETC.

LONDON:

CHAPMAN & HALL, 193 PICCADILLY.

1859.

LONDON:
Printed by G. Barclay, Castle St. Leicester Sq.

PREFACE.

My object in the following pages is to sketch, for the use of the general reader, the history of our Engines of War, so as to enable him to judge of the vast difficulties which have had to be overcome in bringing them to their present state of perfection, and the reasons which, from time to time, cause such an enormous addition to the military expenditure of this country; and as these are subjects to which the public have not generally much time to devote, I have made my narrative as concise as possible, pointing out, at the same time, to those who wish to go deeper into the matter, the writers who will give them correct information.

CONTENTS.

CHAPTER I.

CHAPTER II.

CHAPTER III.

CHAPTER IV.

CHAPTER V.

CHAPTER VI.

CHAPTER VII.

OUR ENGINES OF WAR,

AND

HOW WE GOT TO MAKE THEM.

CHAPTER I.

Of the gradual transformation of the Greek fire of the Byzantines to the gunpowder of modern times, an interesting account has been given to the world by M. Lalanne.* Previous to these researches, the wildest theories had been started on the subject, each of which had found numerous advocates. Some asserted that the secret had been known in the remotest ages by the tribes who dwelt along the banks of the Indus; and amongst others who were of this opinion we find Lord Verulam, in his "Essai of Vicissitudes of Things." Many, again, considered it owed its origin to the Chinese. But during the last century it had

* Mémoires des divers Savants, tom. 1er.

B

become the fashion amongst the *literati* of Europe to trace the discovery of nearly every important invention to China.* Philosophers, after many years profound reflection, having come to the conclusion that there was some reason for believing the original seat of the human race to have been in the East, sought there, not only for the migration of nations, the originality of language, the sciences of husbandry and of architecture, but, dissatisfied with the vicinity of the land of Eden, persisted in going to China for the discovery of logarithms, gunpowder, &c. This China mania had originated with the tales which the first adventurers to that country brought home of its state of civilisation; and when the Jesuits, who ever followed in the track of the Catholic mariner, endeavoured, with their accustomed assiduity, to investigate the annals of the Chinese, that people, who of all nations are the most vain and lying, soon pointed out in their history facts which astonished, if they did not shock, the learned fathers. Their country had been extensively populated before the creation of Adam, and when Noah was repeopling the earth they were a great commercial people; printing they had been acquainted with 1100 B.C., whilst the

* Even of late years we find this belief strong in such men as Schlegel, Philosophy of History; Libri, Hist. des Sciences Mathém. en Italie, vol. i. p. 147, &c.

most moderate of their writers placed their knowledge of gunpowder at 85 A.D. These tales reaching Europe by means of the Propaganda, were eagerly sought after. China became *à la mode*, and, from their porcelain monsters to the able publications of the order of Loyola, everything bearing a cabalistic character was made an object of veneration. Since then, however, the great intercourse which has sprung up with that country has dispelled most of these fables; and amongst others it has been found untenable to maintain that gunnery had been known by them at a very remote period. Still many persist in stating that the Chinese were acquainted with gunpowder centuries before its introduction into Europe, regardless of the fact that nothing had ever been heard respecting this wonderful powder during the long interval which elapsed from the time when the Alexandrian traders first came into personal intercourse with the Chinese merchants at Ceylon, until the followers of Mahomet spread their rule from Thibet and Cashmere to Andaluz, and encouraged science in every capital under their sway. Is it probable that the Arab merchants who carried on so extensive a trade with the most distant countries, and obtained the silks of China in exchange for the rich produce of Venice and Genoa—who, more like princes than mere traders, marched preceded by their banners, and had frequently to fight for their goods,* that they should have remained ignorant of Chinese pyrotechnics, which must have been then considered an extraordinary discovery? For gunpowder is not a simple substance. It is not a mere accidental mixing together of saltpetre, sulphur, and charcoal, but their incorporation into one body in certain defined proportions; and however ill compounded gunpowder may be, the mere forming such a composition shows a certain degree of chemical knowledge. Saltpetre and sulphur must be purified, and no amalgamation whatsoever of these two alone will answer the purpose. The sulphur may be dispensed with, but without some matter which will produce combustion suddenly, at a high degree of temperature, there will be no explosion. Now the only credit which the Arabs gave to the Chinese was skilfulness with their fingers: "The same God who has given a tongue to the Arabians has more nicely fashioned the hands of the Chinese and the heads of the Greeks," says Abulpharagius.† But even Marco Polo or Sir John Mandeville heard nothing on this head in their travels through Chinese Tartary.‡ And at the present

* Don Pascual de Gayangos, Moham. Dynast. in Spain, vol. i. p. 302.

† Ibid. pp. 2, 62, 101, 117.

‡ Those curious in such matters may find some information in an article "On the Origin of Gunpowder," in the United Service Magazine, 1852.

day, the knowledge of the Chinese respecting chemical or physical science is at its lowest ebb.*

Another school of theorists—unbelievers in Chinese science, and who fancied that the introduction of gunnery and the discovery of gunpowder must have been nearly coeval—boldly asserted that it owed its origin to Roger Bacon, because in his "Opus Majus" he thus addresses Pope Clement IV., A.D. 1267 :—"Take saltpetre, charcoal, and sulphur, and you will make thunder and lightning, if you only know the art." But they forgot that Bacon also distinctly affirmed it to be then employed in many parts of the world as a child's toy. The great philosopher of the thirteenth century was no heaven-born magician —no self-taught inventor of magnifying-glasses, gunpowder, and other wonders attributed to him by the ignorant, slothful clergy of his time: he was simply a studious, deep-thinking man, anxious to acquire knowledge, and endeavouring to add his mite to the accumulated science of the past. At an early age he had repaired to the University of Paris, then the most celebrated school in Christendom. But even there, such as were desirous of following the abstruse studies of medicine, chemistry, or mathematics, were obliged to make themselves acquainted with Arabic. For, as if to atone for their sanguinary conquests, the successors

* Appendix A.

of the prophet of Islam had assiduously encouraged learning; and Bagdad, Cairo, Damascus, Alexandria, Fez, Seville, and Cordova, had become frequented by poets, historians, physicians, and philosophers, attracted by the common language which united together the several nations spreading from Asia Minor along the southern shores of the Mediterranean to Andaluz, and which gave far greater freedom of intercourse than the barbarous Latin of Christendom. But if Arabic had become the language of the learned, it must be borne in mind that it did little more than preserve the knowledge of the physical and natural sciences possessed by the Greeks, whom the Arabs ever looked up to as masters on these points.* Now, as far back as the seventh century an incendiary composition of an extraordinary nature, commonly termed "Greek fire," was known to the rulers of Byzantium, and the nature of its composition ordered by them to be kept as a state secret:—"Thou shouldst above all things," wrote the Emperor Constantine Porphyrogenetus to his son,† "direct your cares and your attention to the liquid fire which is

* See Dr. Watson, Elements of Chemistry, vol. i. p. 17; Wotton's Reflections, pp. 121, 133; Pauw, Recherches sur les Egyptiens et les Chinois, liv. i. pp. 376–429. Geber, writing in the ninth century (D'Herbelot, p. 387), confesses he had drawn most of his science from the ancients.

† Imperium Orientale de Banduri, tom. i. p. 64; Meursius, Opp. tom. vi. De Administrando Imperio, c. 13

thrown by means of tubes; and if the secret is dared to be asked of thee, as it has often been of me, thou must refuse and reject this prayer, stating that this fire had been shown and revealed by an angel to the great and holy Christian Emperor Constantine. By this message, and by the angel himself, it was enjoined to him, according to the testimony of our fathers and our forefathers, to prepare this fire but for Christians only, and solely in this imperial city; never to divulge it or allow it to be shown to any nation whatever. And that great Emperor,* to make certain of his successors' secresy, caused to be engraved on the holy table of the Church of God curses against whoever should dare to reveal it to a foreign nation. He decreed that the traitor should be considered as unworthy of the name of Christian, or of any trust or honour: that were he possessed of any dignity, he should be deprived thereof. He proclaimed him anathematised for ever; he declared him infamous, whosoever he might be, emperor, primate, prince, or subject, who should attempt to violate this law." And strictly were these mandates obeyed, for of the very many Christian leaders who passed through Constantinople on their way to Palestine, not one brought back any further information on the subject than an indistinct notion of the terrors inspired by the Greek fire. In the beginning of

* Constantine Pogonatus.

the thirteenth century, however, we find the secret possessed by the Arabs; but whether they obtained it by means of some renegade artizan, or as a bribe to the Sultan of Iconium, who in 1210 A.D. assisted Alexius II. against the Princes of Nicæa, we do not know.* By them it was soon divulged, and in the thirteenth century we find Brother Ferrarius, a Spanish monk and a contemporary of Bacon, in an epistle to one Anselm, which is preserved in the Bodleian at Oxford, enclosing certain receipts translated from the Arabic, amongst which we find:—"To make Greek fire, take xx pounds of saltpetre, viii pounds of sulphur vivum, and v of willow charcoal; reduce them all similarly to powder, and of this powder make what you please. For flying powder, to compound the nitrate powder, take one part of sulphur, two of charcoal of the wood of the willow or lime-tree,

* Under the name of Greek fire the early chroniclers of the Crusades classed every denomination of incendiary composition; but the first who mention the effects attributed to it by the Byzantine historians are Olivier l'Ecolatre, in his account of the siege of Damietta, and Joinville, when they found it used by the Arabs. No better proof, however, can be given of its late introduction into Arab warfare than that Geber, living in the beginning of the ninth century, was acquainted with sulphuric acid; and Avicenna, in the beginning of the eleventh century, mentions several varieties of saltpetre; yet neither say a word respecting any compound of sulphur and saltpetre.—F. Hoefer, Hist. de la Chimie, tom. i. p. 309.

and six of saltpetre, to be very finely ground upon marble or porphyry, after which the powder is placed in a tube at your pleasure, either to be made to fly or to make thunder. . . . . And note that the case for the rocket should be long and thin, and filled with the aforesaid powder, and well crammed in; but the case for making thunder should be short and thick, and half filled with the aforementioned powder, and carefully tied at each extremity with strong thread: and note that in each tube a small aperture should be made to make it flash." And that able antiquarian Sir Francis Palgrave, to whom we are indebted for the discovery of this document,* takes care to remind us how we have unwittingly preserved the origin of Greek fire in our word "cracker," from the Anglo-Norman "le crake," for "le Grec."

To a casual observer it may appear strange that for six centuries nitrate powder should have remained a state secret, and during that period have undergone no manner of improvement. It is, indeed, worthy of deep consideration how Providence regulates the knowledge which we possess in accordance with the social state of mankind. In the seventh century Constantinople was the only enlightened portion of the world; and though the provinces which had been sub-

* Preface to his Merchant and Friar.

jected to Roman rule were still partially civilised, yet, overrun as they were by northern barbarians, they were rapidly sinking to a level with their ignorant conquerors. Religion had not yet sufficiently regulated the passions of the mass, to permit the introduction of gunpowder as a general means of warfare. Indeed, there was not even that feeling of right and wrong which Christianity has endeavoured to inculcate, to restrain man from turning it into a mere ruthless agent of human destruction, and which in course of time made him look with horror upon a St. Bartholomew or a Gunpowder Plot. As it was, the ingredients of which gunpowder is composed were but roughly freed from their impurities, and merely mixed together on a slab; so that even in its loose state the badness of the material prevented any very rapid combustion. Yet it is wonderful that so much had been found out at a time when, in the Mahommedan as well as the Christian world, schools were confined to cathedrals and monasteries; when libraries were scarce, and works on the more abstruse sciences still more so; and when the chemical student feared to give his discoveries to the world, lest he should be made a martyr.

When first originated the idea of expelling missiles from tubes by means of the sudden explosion of the nitrate powder will probably never be known with certainty, but it must have been shortly subsequent to A.D. 1320. That it was not previous to that date we learn from the silence on the subject of Egidio Colonna and Sanuto. The former was tutor to Philip le Bel, and made the science of warfare his peculiar study, until his death in 1316; the latter, of the noble Venetian family of that name, minutely describes the war engines about the year 1321. Within a very few years, however, after this, cannon of small size, and formed of two tubes, the smaller being termed

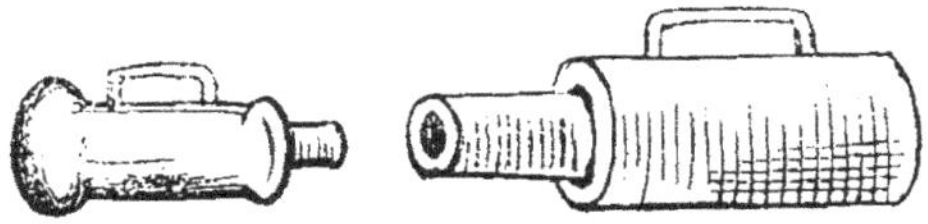

From the Santini MSS. See App. B.

a chamber, in which was put the powder, and which was fixed into the larger one either with a screw or pin, were in use in nearly every European state. A Provisione of the Republic of Florence grants, A.D. 1326, to the priors, the gonfalonieri, and twelve elders, the power to appoint two officers, entrusted with making iron shot and metal cannon for the defence of the castles and villages of the Republic.* In 1338

* This document was first pointed out by M. Libri, in his Histoire des Sciences Mathém. en Italie, tom. iv. p. 487, and was verified by M. de Mas-Latrie of the École des Chartes.

A.D., amongst the stores of the hulk "Christopher of the Tower of London," were three iron cannon with five chambers, and in the barge "Marie de la Tour" one iron gun with two chambers, and another of brass with one chamber.* The earliest French record is also of the same year;† but in both these countries these memorandums were merely accounts rendered of government stores relative to the English invasion, and not details of a recent invention;‡ and we may therefore believe Barbour was correct when he says with respect to Scotland, that

> "The wss of thaim had nocht been sene."§

The extensive trade which the Italian states carried on with the Mahommedan powers would alone account for their becoming acquainted with gunnery before the north of Europe; but it is remarkable that in Spain, Italy, France, England, and Germany, the dates of the public documents in which cannon are first mentioned scarcely differ from one another by ten years, a fact which could

* Hist. Royal Navy, by Sir Harris Nicolas, App. vol. ii.

† Lacabane, De la Poudre à Canon et de son Introduction en France. Paris, 1845.

‡ See especially Collection des Documents inédits, Archives d'Administration de la Ville de Reims, tom. ii. seconde partie, pp. 786, 791.

§ The Bruce, B. 11, v. 948.

only have arisen from the manner in which armies were raised in those days. A certain portion consisted of the feudatories of the country whose name it bore, but by far the greater part were mere mercenaries and adventurers, attracted by the prospect of pay and plunder; and who, as they roved from one service to another, carried with them the fame of such as were skilful in smithery, or cunning in devising engines of destruction. This gradual yet steady diffusion of the knowledge of the power of gunpowder is also curiously exemplified in a memorandum found amongst the archives of Tournay, by which it appears that, in 1346 A.D., the town council, then ignorant of gunnery, having heard that one Peter of Bruges, a worker in tin, knew how to make certain engines named "connoilles," by means of which he could fire projectiles into a town, sent for him, and ordered him to make one for their inspection, promising that, if successful, he should receive an extensive order. Peter made one and took it outside the Moriel Gate, and fired from it a quarrel, having a leaden head, weighing two pounds. No one saw where the bolt fell, but it was soon discovered that it had killed a man beyond the second wall. Alarmed at this accident, Peter took refuge in a sanctuary, but the town council, taking into consideration the facts of the case, granted him a free pardon.* Thus from city

* Mémoires de l'Académie de Bruxelles.

to city travelled the news of this mighty invention: every petty state, every independent baron anxiously endeavouring to obtain some of these extraordinary engines, until, very few years subsequent to their first introduction, Petrarch could write, "I am astonished that thou also dost not possess those brazen globes, which, impelled by fire, start off with a horrible noise. Was it not sufficient that the anger of an immortal God should thunder from the sky, but must a fragile being thunder on the earth? This scourge was, till lately, so scarce, as to be looked upon as a prodigy; but now that men's minds are apt for the most wicked deeds, it has become common, and is made use of as much as any other weapon."*

From Valturius.

It was, indeed, to prove a scourge to those petty princes who for so many centuries had ruled their serfs with a rod of iron. Rapidly did the mechanic grow into importance, when from every smithery resounded the hammer which welded the hand-gun, by means of

* De Remediis utriusque Fortunæ.

which the villein was placed on a level with the knight; and the middle classes began not only to be tolerated, but even to be respected, when suzerains were glad to avail themselves of the aid in gunnery which, from its expense, their goodly towns could alone furnish them with.* From this epoch dates the decline of the old feudal system in Europe, at least that stern tragic portion of it which had created the necessity of every man's house being his castle, a state of society difficult to picture at the present day; when the royal signet no longer protected the subject, and that the burgess found the crenelated wall a far better proof of his charter than any parchment could afford him; when France possessed its ten thousand walled towns and fifty thousand fortified castles; when

From the Santini MSS. See App. B.

* In 1382, when the men of Ghent attacked the Earl of Flanders, they took into the field two hundred chariots armed with guns.—Froissart, liv. ii. chap. cliv. In 1383, Sept. 13, Charles VI. of France thanked the citizens of Tournay for their assistance with guns and gunners. Do. again, 1386.—Collection des Documents inédits concernant l'Histoire de la Belgique, par L. P. Gachart. Bruxelles, 1853.

in Italy and Germany almost every house was a citadel, a perfect lazaretto, which fear and suspicion caused to be watched by day and night; when the abodes of religion itself were not exempt from the universal lawlessness, and that many a monastery, though within a fortress, was surrounded with lofty walls, flanked with bastion and turret, and could be approached but by drawbridge and crenelated gateway.* England felt the change, perhaps, less than any other nation. Few of its cities had ever seen a foreign foe within their walls, and both kings and barons had too often required the assistance of the commons against one another to think of uniting to crush them. The middle classes had therefore gone on progressing in wealth and power, cheerfully contributing to the wants of their monarchs, in exchange for their respected privileges. Besides, unless we except the reign of Stephen, no man was allowed to erect fortifications round his dwelling without royal license, of which our monarchs were at all times particularly chary.† And already were applications for such

* Du Sommerard, "Arts du Moyen Age," vol. i.; Augustin Thierry, "Concernant la Révolte de Vezelay," in his "Lettres sur l'Histoire;" Monteil, "Hist. des Français des divers États," liv. i. p. 131.

† Thus Edward I., in 1294, grants to Hugh de Frene leave to fortify and crenelate his dwelling at Mockes, Co. Hereford, with a wall of stone and lime, *without tower or turret,*

a purpose falling into disuse under the vigorous rule of Edward III., except in those parts more particularly subject to the inroads of the Welsh and Scotch. Indeed, many of the leading nobility had erected spacious mansions within their county towns; amongst others, the palaces of the Bishop of Ely in Holborn, and of the Earl of Lancaster at Lincoln, remained up to a late period examples of the deference which, even in those days, was paid to the laws in England.*

We should, however, greatly err were we to suppose that gunpowder, on its first introduction, immediately superseded all other kinds of military engines: far from it; it formed almost only a nominal portion of the engineer's resources, and would seem to have possessed far more of a moral than a physical effect. Under the names of "trebuchet," "biffa," "mangonel," "pierrier," "bombarda,"† &c., we find the powerful sling, the

so that the wall under the crenelation be 10 *ft. high.*—Rot. Pat. 21 Ed. I.

* Turner, Domestic Architecture; Archæologia, vol. i. p. 362; Archit. Ant. of Great Britain, by J. Britton, vol. ii. p. 79.

† For explanation of these terms see Ducange's Glossarium. The Emperor Louis Napoleon, in his Études sur l'Artillerie, enters fully and ably on this subject. They were made use of up to the sixteenth century; and Ramelli, engineer to Henry III. of France, in his Diverse et Artificiose Machine, Paris, 1588, gives a drawing of a formidable "trebuchet."

pao of the Chinese,—machines from which were

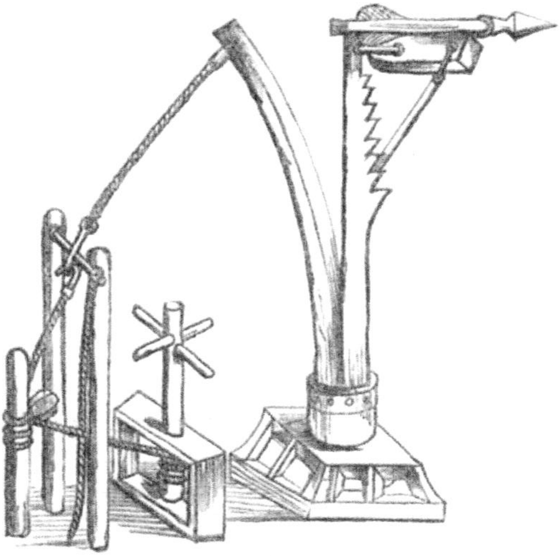

Arrow Fired at a Siege, from Valturius.

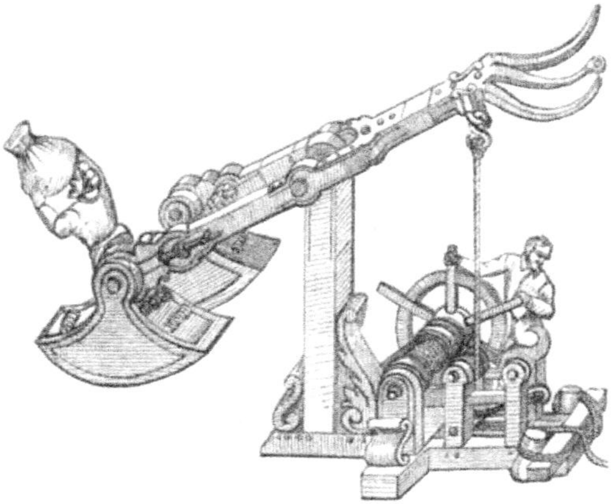

Mangonel, Sixteenth Century, from Ramelli.

cast not only huge stones, but also iron pots filled

with inflammable matter, portions of putrefied carcasses, &c.; and an Earl of Derby, at the siege of

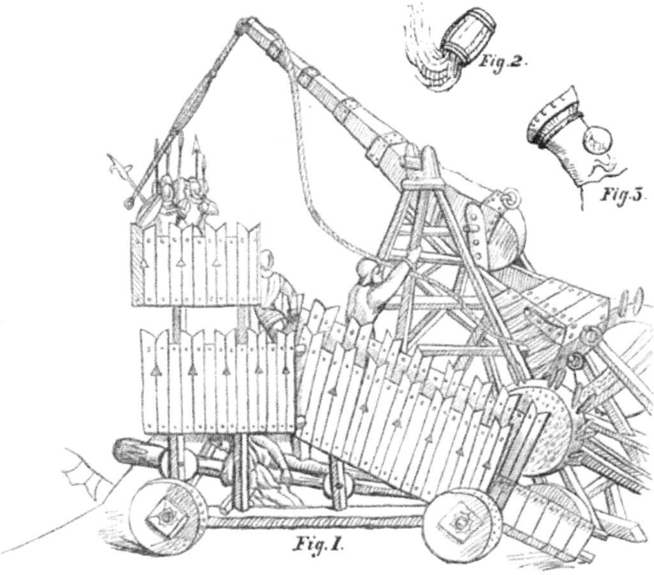

Sow and Mangonel, Fifteenth Century, from the Santini MSS.

Auberoche, in 1345, affords the singular example of thus returning a messenger who had been sent to him by the townspeople to demand terms.* Enormous arrows also were fired from cross-bows, often thirty feet in length, drawn on wheels, and which, according to the manner in which they were wound up, were known as "arbalestes à tour," "à croc," "à tillol," "de passe," "quar-

* Froissart, Chron.

tier," "espringole," "ribaudequin," "martinez," "sougnies," &c.* The difficulty, however, of managing these engines directed the attention of engineers towards improving gunnery; the simplicity of which promised important results, as castings became more perfect and powder improved.

* Colonel Dufour, of the Swiss army, had one of these large cross-bows constructed, which carried a four-pound quarrel eight hundred yards.—Louis Napoleon Bonaparte, Etudes sur l'Artillerie.

CHAPTER II.

The Emperor Louis Napoleon, in his Études sur l'Artillerie, has revived the discussion respecting whether artillery was at first introduced as a field-weapon or confined to sieges, and supports the statement of Villani and of the "Grandes Chroniques de St. Denis," by an extract from a MS. Froissart preserved in the library of Amiens, wherein it is stated that "the English let off certain cannon, which they had in the action, *to alarm** the Genoese." But this solitary example of a single MS. is by no means a proof to be relied on, for copyists were constantly in the habit of making additions of their own; and as no other copy has been found to corroborate this, it is most probably an interpollation. It is, however, surprising that Froissart's statement should ever have been opposed to Villani's. The former was not born till 1337 A.D., and did not begin his Chronicle till 1356, whilst Villani, at the time of the battle of Cressy, was an aged man, well acquainted with courts and

* The French word *esbahir* well expresses the astonishment as well as fright caused by those engines to the ignorant.

camps; but, as the Emperor very judiciously observes, these guns could only be compared to three

Gun and Quarrel. Temp. Edw. III. Sloane MSS. 795.

very inferior muskets, and could have had nothing to do with putting 50,000 men to flight. Such controversies are curious, however, as demonstrating the fact that, possessing, as we do, inexhaustible iron mines, diffusing wealth and comfort throughout the land, and yielding a metal, the working of which has been brought to a singular perfection, we cannot realise the great cost of iron and brass in the fourteenth century, or the difficulties encountered in making the gun sufficiently powerful to stand the shock of explosion of even the powder of that day. Now so scarce was iron at the time of Edward the Third's invasion of France, that "it was accorded and established, that iron made in England, and iron brought into England, and there sold, shall not be carried out of the said realm of England, upon pain of forfeiting double to the King; and the justices assigned shall have power to enquire of workmen, and the other justices whom the King will thereto assign, shall have power to enquire of them that sell the iron at too dear a price, and to punish them after the quantity of the trespass."* And so limited was the knowledge of casting, that the cannon of this period were little larger than our duck-gun. In

* Statutes of the Realm, 28 Ed. III.

1339 A.D., we find ten guns made for Cambray, five of iron and five of brass, at a cost of 25 liv. 2*s*. 7*d*., which, at the price of iron of the time,* allows but 46lbs. weight to each gun, supposing all to have been of iron; and as five were of brass, the weight must have been much less. Such engines could therefore have been of no use against masonry, and consequently Edward III., in his expedition against Calais, out of an engineer corps of 366 men, took but six gunners.† Besides, saltpetre and sulphur were also rare and expensive materials. These guns for Cambray had been delivered in October 1339, yet it was not till December of that year that the materials for powder were given, and then only to the value of 11 liv. iiis. iiid., tournois.‡ From a receipt also of Thomas de Ruldeston, who, as Keeper of the King's Wardrobe at the Tower, had charge of all engines of war, we find that on the 4th May, 1346, he had received from John Cook, Clerk of the King's Great Wardrobe, by the hands of William de Staines, for his guns, 912lbs. of saltpetre, and 866lbs. of sulphur. In the month of November following, the King issued a writ commanding that all the saltpetre

* Lacabane, De la Poudre à Canon et de son Introduction en France. Paris, 1845.

† Harleian MSS., No. 782, fol. 63.

‡ Lacabane.

and sulphur which could be found should be sold to Ruldeston for the King's use. It amounted to only 750lbs. of saltpetre, and 310lbs. of sulphur. In the year following De Ruldeston paid 167*l*. 2*s*. 2*d*. for 2021lbs. of saltpetre and 466lbs. of sulphur, being 1*s*. 6*d*. per lb. for saltpetre, and 8*d*. for sulphur; money being then worth about fifteen times the value of what it is at present.*

It is not, however, until about the middle of the fourteenth century that gunnery may be said to have begun its real existence, at which period one Berthold Schwartz appears to have discovered the granulation of gunpowder, and the consequent increased force of the explosion necessitated greater thickness in the metal, which caused great drain on the treasury. Indeed John, king of France, was obliged to order his "Généraux des monnoies" to diligently inquire into the quantity of brass to be found in France, "for the purpose of making such artillery, and to prevent its being taken out of the country."† Henceforward we find the "great

* Archæologia, A.D. 1847, p. 379. Ruding, Annals of the Coinage of Great Britain.

† Lacabane, De l'Introduction de la Poudre en France. Paris, 1845. Schwartz' discovery has been handed down to posterity by a picture preserved in the degl' Uffizi Museum, at Florence, painted by Giacomo Crespi, at the end of the fifteenth century, which represents the German chemist in his laboratory, where, surrounded by several of his apprentices, he is working at the manufactory of gunpowder in a mortar, upon

gunnes" of the middle ages. Yet many were the trials made by the bronze-founders before they succeeded in obtaining a metal sufficiently tenacious to resist the improved "gonne-poudre;" but, nothing daunted, every form and size was successively tried, and great was the successful artist's reward; for these huge engines were of extreme value, and worthy of a monarch's gift,* although their usual fate was, either to be abandoned, owing to their impeding a march, or to burst after a few rounds. Thus, in 1415 A.D., Gerard Sprong

Santini MSS. Fifteenth Century, Early Part.

which is inscribed "Pulvis excogitatus, 1354; Dania Berthold Schwartz."

* As late as 1510 A.D., James V. of Scotland gave liberal possessions to Robert Borthwicke, a noted artificer, for making certain pieces of ordnance for Edinburgh Castle.—Hollingshed; Hist. Scot. From the Pell Records we find that Henry IV. of England was fond of experiments of this nature:—"To Simon Fleet, Esq., Keeper of the King's Wardrobe in the Tower of London, in money paid to his own hand for making and constructing a certain great cannon, newly invented by the Lord the King himself, 25*l.* 6*s.* 8*d.*—Pell Records, 9 Hen. IV. Gachart, in his Documents concernant l'Histoire de la Belgique, part i. p. 246, publishes an interesting letter of Philip, duke of Burgundy, 1453, respecting a present of ordnance to the Knights of Rhodes.

petitioned Henry V. that a warrant might be issued, commanding the Treasurer and Barons of the Exchequer to grant him a discharge for the metal of a brass cannon called "Messenger," weighing 4480lbs., which burst at the siege of Aberystwith; of a cannon called "Kynge's daughter," burst at the siege of Harlech; and of a cannon which burst in proving.* We find also that the Italians, who prided themselves on their knowledge of bronze-casting, were not much more successful. "I inform your lordships," writes Duke Federigo of Urbino to the citizens of Sienna, "that we cannot move from hence, because the Marzochesca bombard has not yet been removed, and I have not had it broken up, because the Messrs. Borghese tell me your lordships wish for it; and the body of the last bombard which burst is still here, for its carriage broke down on the march, as also its chamber, and was left by the way."† Govern-

* Proceedings of Privy Council, vol. ii. p. 339. Curiously enough, Lydgate, in his contemporary poem of the Siege of Harfleur, gives us the calibre of these pieces. "Messenger" threw stone-shot 30lb. in weight, and "Kynge's Daughter" 45lb. From the Issue Rolls of the Exchequer, 1 Henry IV., quarrel gunnes were *7s.* each, saltpetre *4d.* per lb., sulphur $4\frac{1}{2}d.$ per lb. In 35 Henry VI. saltpetre had risen from *5d.* to *7d.* per lb.; iron was *3d.* per lb.—Issue Roll, Easter, 35 Henry VI., 19th May, 6th July.

† Denistoun, Dukes of Urbino.

ments had been obliged to resort to brass for their ordnance, on the discovery of granulated powder, on account of the inferiority of cast-iron;* and experiments were made with a view of reducing this expense, by substituting forged iron tubes, strengthened with iron hoops and bands; but even this was not very economical, as we find by an account of 3 Henry V., where was paid to John Stevens, of Bristol, for the making a great cannon, as well as for iron, coal, and timber purchased by him, as for divers other expenses incurred by him on account of the cannon aforesaid, 107*l.* 10*s.* 8*d.*,† equal to about 1000*l.* of our money.

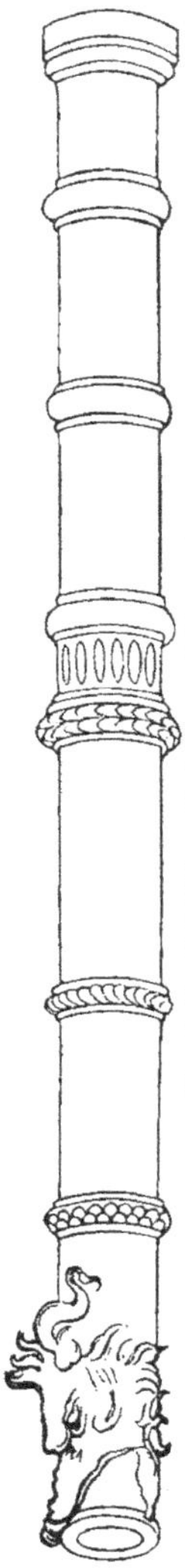

Giorgio Martini, Fifteenth Century; latter part.

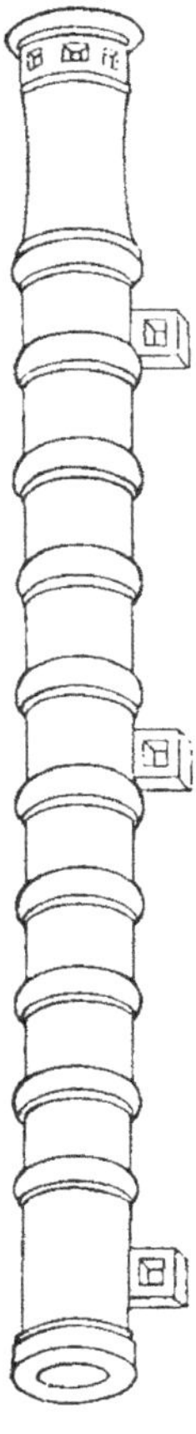

Giorgio Martini, Fifteenth Century; latter part.

In spite of these difficulties, we find Henry V. taking with him

* For the early use of cast-iron in cannon, see Lacabane, De l'Introduction de la Poudre en France.

† Issue Rolls, Henry V., p. 332.

to France three master-gunners, twenty-five gunners, and fifty assistant gunners,* showing the

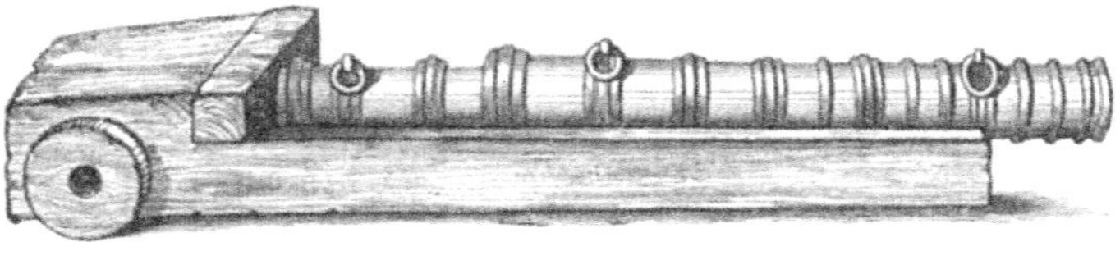

From the Wreck of the Mary Rose, sunk temp. Henry VIII.

steady progress gunnery had made since its introduction into warfare, and the wars of the House of Lancaster in France forced the French to acknowledge the advantages to be derived from it; consequently under the able rule of Charles VII. and Louis XI. they organised a very fine artillery, and the power derived therefrom was amply shown in the success which attended the invasion of Italy by Charles VIII. The mere rumours which spread throughout that peninsula respecting his artillery completely overawed its inhabitants; and, though grossly exaggerated, his means were certainly wonderful for that age, and the practical effect of his guns far surpassed anything that had been said of them. The Italians still used the heavy cumbersome bombards of iron or bronze, which, drawn with difficulty, were loaded with still greater, and thus,

* The list will be found in Sir H. Nicolas' Agincourt, p. 97 of Appendix.

in battering a town, gave the besieged plenty of time to erect fresh works in rear of those destroyed. But the French, never making use of any but bronze cannon, having replaced stone-shot with iron of much smaller calibre, but of equal weight, drawing their pieces with horses instead of oxen, and possessing experienced gunners, could keep up with the rest of the army, and, throwing up their batteries with rapidity, could cause in a few hours the same effect as the Italians could in as many days. "Whence all Italy was filled with dread of Charles' army."* Yet were the Italians devoid neither of courage nor military capacity, and Francesco Sforza, Roberto di San-Severino, Frederigo di Montifeltro, Fabriccio Colonna, Giovanni Jiacome, Trivulzio, &c., were as brave leaders as any in Europe. But to accumulate large stores of military engines, and at the same time introduce these improved weapons, required means which the small principalities could not afford. The Italian soldier had, therefore, in his intestine

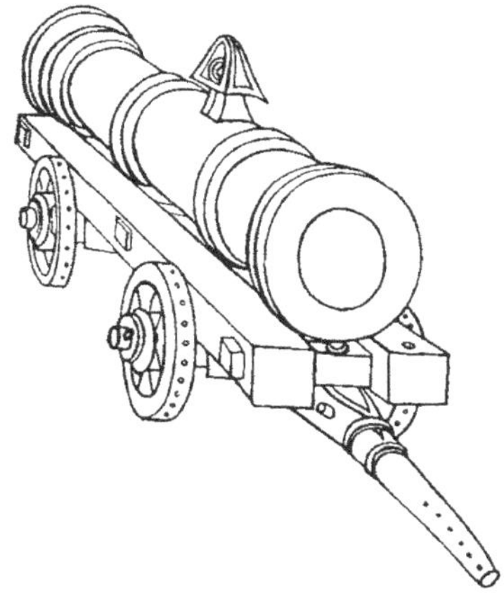

From Valturius.

* Guichiardini, liv. i. chap. 18.

wars, learned to substitute craft to boldness. If he besieged a town, he usually encamped close outside the walls and raised a few earthen works, strongly palisaded, which prevented all egress, and trusted to the effects of famine and disease, which he knew must soon break out in the city. Wherefore we find, in their military writers of that age, much good counsel respecting the site which should be chosen for a fortress: that it should be situated at a distance from any marsh, so that the air round about should be pure; that water should be plentiful; that it should be able to grow its own supplies; and many other points relative to health and comfort, which were of far more importance to them than the solidity of their walls, or the manufacturing of their engines of war.* This system of warfare they even carried into the field.† But the facility with which the French battered down their walls, and the fearful slaughter they made of the inhabitants of every town they took by assault, induced the Italians to turn their fertile brains to some invention by which they could in some way obviate such terrible disasters.

* Battista Alberti. Francesco di Giorgio Martini, edited by C. Promis. Turin, 1841.

† Sismondi, Républiques Italiennes. See especially the campaign of the Marquis of Mantua against Gonzalvo di Cordova.

The first system of modern fortification with which we are acquainted is the simple wall, with

Clermont, Fourteenth Century. From Du Sommerard, "Arts du Moyen Age."

towers at intervals,—a system which answered perfectly at a time when the only weapons were the sword and arrow. This appears to have prevailed up to the time of the first Crusades, when the introduction of the sow* and trebuchet from the East rendered it necessary to form a ditch outside the wall, an idea which does not seem to have been introduced into England much before the time of John; † and, later, the wall itself was strengthened with machicoulis. As the power of the trebuchet increased, the walls and turrets had to be raised and thickened; and this being a great expense, the Norman gentleman allowed his dungeon to fall into decay, whilst the monarch increased his power by erecting those magnificent fortresses, of which Conway Castle is so fine an example. The citizens, however, still maintained their privileges of contributing towards the defence of their towns—a tax which they would force the wealthy monastic institutions, much against their will, to join in.‡ As gunnery became perfected, the King's castle—in spite of its buttresses and towers—could offer no sufficient resistance, and town walls became

* So called from its enabling the besiegers to burrow at the wall under protection of its hide cover.

† For instance, the citizens of London, as an additional security to the city, began to encompass the wall with a spacious and wide ditch.—Stow, Sur. London.

‡ See Recueil des Monuments inédits de l'Histoire du Tiers-Etat, tom. i. pp. 573 and 733; De Cheruel, Histoire de Rouen.

Castello Vecchio, Florence.

useless. Then huge Bastilles were erected, large, square, gloomy buildings, with walls thirty feet in thickness. But to erect these was not only costly, but also a matter of time; and in the course of time it was discovered by monarchs that these citadels might prove as effectual in the hands of their subjects as they did in their own, and the skill of the engineer was redirected towards the strengthening of cities. Earthen mounds, called "ravelins," were erected outside each gateway, to prevent the guns being run up, as they were accus-

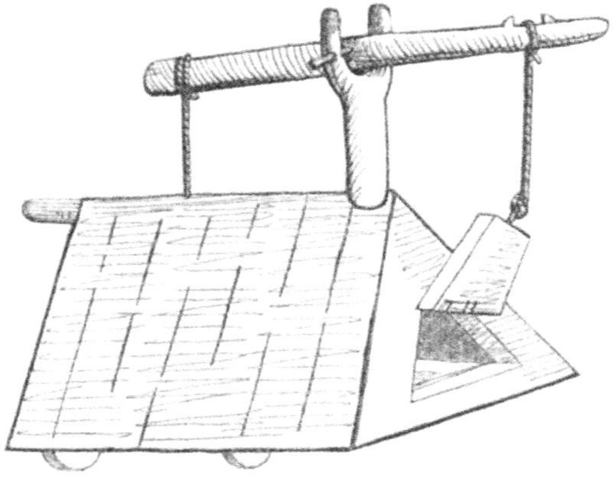

From the Santini MSS. Mantelet for running up Guns under the Walls.

tomed to be, under protection of a covering; and subsequently *boulevards*, or bulwarks, also earthen mounds, were thrown up within the city against the walls, for the double purpose of strengthening them, and raising guns sufficiently high to fire over them.*

Such appears to have been the prevailing system of fortification when Charles VIII. entered Italy. But as town after town, and fortress after fortress, fell into his hands, the Italian engineer put his wits to work to endeavour to keep up the superiority of defence over offence. The earthen parapet erected within the walls being found to facilitate an escalade when the wall was breached, a strong convex rampart was formed within each bastion, (of stout timber and earth,) guns being

Giorgio Martini, Fifteenth Century.

placed in casemates at the bottom, so as to command the ditch of the rampart. Then

* Della Fortificazione della Citta, di M. G. Maggi e del Capitan T. Castriotto. Venezia, 1564.

counter-forts, or buttresses, were erected some twenty yards from one another, so that the bulwark might support itself even when the artillery had destroyed the wall in front; and the ravelin being found to act as a shield to a fortress, it was recognised that it should be defended by the fortress, as well as defending it, and that, therefore, it should be placed at a distance sufficient to be supported in flank.*

But for the next fifty years Italy swarmed with engineers, amongst whom we find such men as Antonio di San Gallo, Michele Angelo, Leonardo da Vinci, and San Michele; and as it was an age in which there was far more practice than theory, the art of fortification made rapid progress. Few plans were executed without several engineers of eminence having previously been consulted. Some advised one thing and some another;† and as men then lived and died in harness, the advantages to be derived from each particular suggestion were soon tested. The improvements in gunnery had caused the diameter of the tower to be increased, and a round form to be preferred to the square; but the former not affording sufficient defence to the flank

* Trattato di Architettura Civile e Militare de Francesco di Giorgio Martini. Turin, 1841, by Carlo Promis. Born in 1433 at Sienna, he was still alive in 1503. MSS. of Leonardo da Vinci. B., fol. 16, 95; N., fol. 42, 47.

† See Life of Antonio di San Gallo, in Vasari.

on each side, it was gradually reduced to a semi-circumference, and then to an angle. Their

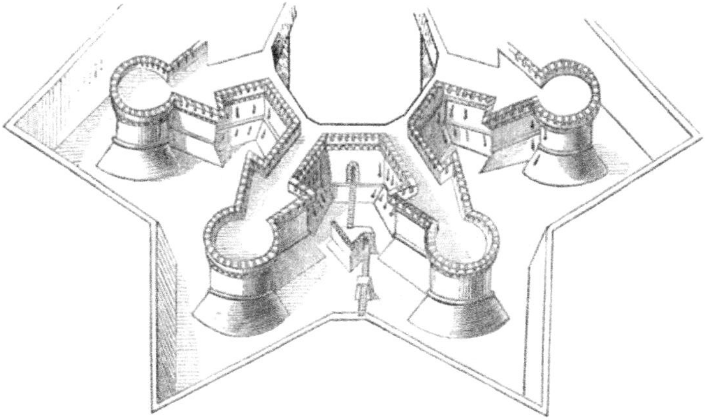

Giorgio Martini, Fifteenth Century.

distances also from one another gradually increased with their cost; indeed, some engineers were satisfied with their being flanked by cannon shot, but experience demonstrated in time that it should be by the harquebuss. For a length of time also the space of wall between them, called the "curtain," was not strengthened with bulwarks, on which guns might be placed, until their necessity was practically felt; and so it was with the height, and depth, and slope of the ditch; the amount of masonry to be used in the walls; the height, thickness, and distance of counter-forts, &c. For, like all other great changes in this world, the modern system of fortification has been the result of time.*

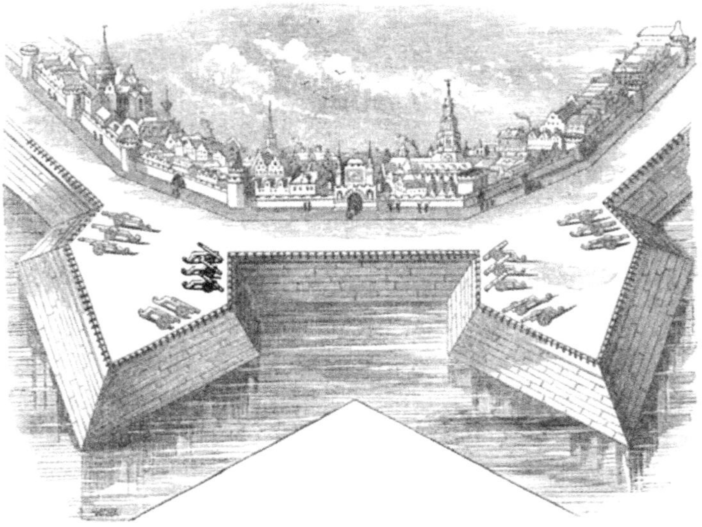

Marchi, Sixteenth Century.

These innovations in the art of war aroused the attention of most governments, but the English remained satisfied with patching up their fortresses according to the good old custom,† a subject which

* The Italians are particularly rich in illustrations of the masters of this period. See especially Le Fabriche Civili, Ecclesiastichi, e Militari, di Michele San Micheli, disegnata e incisa da Ronzani Francesco e Lucioli Girolamo: Venezia, Giuseppe Antonelli, Editore. 1832. Francesco de Marchi, Archittetura Militare, 1546. A very fine edition was published at Rome in 1810.

† See the orders for the repairs of Calais fortifications in 24 Hen. VIII., in MSS. Cotton, Calig. E. III. pp. 77*b*, and 78.

it is curious to find lamented by a boy so young as Edward VI. Writing to his favourite, Barnaby Fitz Patrick, chieftain of Ossory, he thus describes Portsmouth:—"We went to Portsmouth toune, and there viewed not only the toune itself, and the haven, but also divers bulwarkes, as Chatertone, Haselford, with others. In viewing of wich we find the bulwarkes chargeable, massie, wel rampared, but il favored, il flanked, and set in unmete places, the toune weaklie in comparison of that it ought to be, to houge great (for within the walles are faire and large closes, and much vacant rome), the haven notable, great, and standing by nature easie to be fortified. And for the more strength thereof we have devised two strong castles on either side of the haven, at the mouth thereof."*

* Camden, Antiq., edited by Gough, 1806, Hantshire.

CHAPTER III.

WHEN taking into account the sudden expansion of military science throughout Italy, at the commencement of the sixteenth century, it must be borne in mind how much its military engineers of that period were indebted to the emigration of the Greeks after the fall of Byzantium, and to the discovery of printing, which placed within the reach of all the literary treasures which had been imported. For science, like all other human crafts, can only progress by those who follow it imparting their ideas to one another, and thus correcting errors, or adding to the general store of human knowledge; indeed, it is thus only that we could account for the extraordinary theories which are found in such profusion amongst the highly-prized note-books of one of these, whom we shall take for our guide—Leonardo da Vinci.* As a young man, the sciences of artillery and of fortification seem to have been his special studies.

* Appendix C.

" Most illustrious Lord," wrote he to Ludovico il Moro, "having now sufficiently seen and considered the works of those who repute themselves masters and inventors of war-like instruments, whose operations and inventions I have found to be nothing out of the common way, I offer, without wishing to derogate from the merits of others, to make known to your Excellency the secrets I possess; and I hope to be able to give sufficient proofs of my capability at any time you may be pleased to appoint, and in any of the things which I now propose, which, for the sake of brevity, shall be here underwritten.

"1. I understand the construction of light bridges, easy of carriage, and equally adapted to pursue or escape from an enemy, secure from fire, and as easy to remove as to replace; and also the means of destroying those of the enemy.

"2. I understand how to cut off the water from the ditches of a besieged place, and how to make bridge scaling-ladders, and other implements necessary on such expeditions.

"3. *Item.* If, on account of the height or inaccessibility of a fortress, it cannot be effectually bombarded, I have the means of destroying any stronghold (*roccia*), or other fortress, if its foundations are not on rock.

"4. I possesss also a kind of bombard, portable and easy to carry, and with which to throw hail-shot (*minuti di tempesta*), and with the fire of which I can cause great terror to the enemy, with great loss and confusion.

"5. *Item.* I have a way to mine both narrow ways and round about (*distorte facte*), without any noise, so as to reach a certain . . . . . which may be required to pass under a ditch or any river.

"6. *Item.* I can make covered chariots,* safe, and not

* Alluding to the small carts, upon which several wall-pieces were placed in battery, much used at that time.

to be damaged, which, entering into the midst of an enemy, with its artillery, there is no body of men-at-arms, however

Temp. Hen. VIII.

numerous, but which will be defeated; and behind these infantry can follow quite safe (*inlesi for illisi*), and without any impediment.

Leonardo da Vinci.

"7. *Item.* In case of need I can make bombards, mortars, and passavolantes,* of beautiful and useful shapes, out of the usual manner.

"8. Where the effect of bombards fail, I can make briccoles, mangonels, trebuchets, and other machines of marvellous efficacity, and out of the usual way; and to sum up, according to the nature of the case, I can invent various and infinite things to injure.

"9. In relation to naval affairs, I can construct many most ingenious instruments for offence and defence; and the ships that I will make can resist the shot of the greatest bombard, and powder or smoke.

"10. In time of peace I think I can bear comparison with any one in architecture and compositions of houses, both public and private; and in conducting water from one place to the other.

"*Item.* I understand the different modes of sculpture, either in marble, bronze, or terra cotta. In painting I also think myself equal to any one, let him be who he may.

"I could also execute to your satisfaction the bronze equestrian statue which you propose to raise to the memory and lasting honour of my lord your father and the renowned house of Sforza.

"Should any of the above-mentioned things appear either impossible or improbable, I am willing to prove the truth of what I say, by making experiments of the same in your arsenal, or in whatsoever place may please your Excellency, to whom I most humbly recommend myself." †

* The "Flying past," name of an Italian piece of ordnance, of about 16 lbs. shot.

† This letter is in the Ambrosian Library. A very faulty translation is to be found in a very careless work, entitled "Life of Leonardo da Vinci, with a Critical Account of his Works," by J. W. Brown. London, 1828.

The Sforza having accepted Da Vinci's services, he remained at Milan until its occupation by the French, when he served for a short period in the post of chief military engineer to Cæsar Borgia.* His knowledge of military engineering was indeed wonderful, when we reflect on the works published on this subject, not only during his time, but more than a century later, by such men as Ufano and Collado. Basing all his inquiries upon sound mathematical principles, the Tuscan painter was never satisfied respecting any proposition, unless verified by experience; for, as he says, "I will treat of this subject, but first of all I will make some experiments, because my intention is to quote experience, and then to show why bodies are forced to act in a certain manner. This is the system which should be observed in the study of the phenomena of nature." A sound reasoning, which was many years after given by Tycho to Kepler.† We thus find him pursuing his researches, from the simplest question in geometry and statics to some of the deepest laws of dynamics, especially those affecting the flight of

* A Letter dated 1502, addressed by Cæsar Borgia to the governors of his castles, to give Da Vinci every assistance in his inspection of them, is printed in "Amoretti, Memorie Storiche, &c., di Leonardo da Vinci."

† Venturi, Essai sur les Ouvrages Physico-Mathém., de Leonard da Vinci.—Brewster's Martyrs of Science, p. 211.

projectiles, a theory then new amongst mathematicians; and of all problems it is, perhaps, the one which involves the greatest amount of difficulties, including, as it does, the several points of velocity, of friction, resistance of the atmosphere, motion, gravity, theory of curves, ricochet, and, when applied to practical gunnery, the form and nature of the projectile, the strength of gunpowder, and the tension of metals. Yet in all and each of these points we find him a century in advance of his contemporaries.

At the beginning of the sixteenth century the Aristotelian philosophy was still in all its vigour; and if the question was asked, why a stone which had been thrown continued to move a certain time after it had left the hand? it would be stated, in reply, that the air which followed it in rear continued to impress it with motion. Few men had as yet suspected that motion was rectilinear by its nature, and that it would continue to be so, and with an equal velocity, if not counteracted by some obstacle. On the contrary, it was generally supposed that there were certain motions circular by their nature, and which, according to Aristotle, were the only unchangeable ones. There were also certain rectilinear motions, arising from a peculiar *appetitus* of certain bodies to reunite at the centre of the universe, or to fly from it, which constituted weight or lightness. Motion also was divided into natural and violent. The former arising from the nature itself of a body, such as the circular motion of the planets, and that of heavy bodies; the other was so contrary to the nature of bodies, that it could not last any time without a constant renewal of the motal force, such as a stone thrown. As to the ideas of the age respecting the theory of projectiles, they were based on errors. Many, such as Daniel Santbach,* imagined that a body urged on with violence described a straight line, and then that it fell perpendicularly; others thought that a shot, on leaving the muzzle, described a straight line; that after a certain period its motion grew slower, and then it described a curve, caused by the forces of projection and of gravity; finally, that it fell perpendicularly. Tartaglia,† again, seems to have originated the notion that the part of the curve which joined the oblique line to the perpendicular was the arc of a circle tangent to one and the other, a principle as erroneous as those preceding it; and it was only by an accident he nearly stumbled upon one truth in the theory of projectiles, when he stated that the greatest range obtained by equal forces is at 45°. In coming to this conclusion he had reasoned much about the same way

* Prob. Astron. 6, A.D. 1561.

† La Nuova Scientia di Nicolo Tartaglia; also his Quesiti ad Invenzioni Diverse.

as Carden had done about the inclined plane. Calculating that at the angle of 0° the trajectory was null, that by raising the trajectory the range increased up to a certain point, afterwards diminished, and finally became 0° again, when the projective force acted perpendicularly, he concluded that the greatest range must be a medium between these two points, and consequently at 45°.* But Leonardo da Vinci, long before the birth of Tartaglia, taking as his motto, "Science belongs to the captain, practice to the soldier," † boldly asks, "If a bombard throws various distances with various elevations (*archi di moto*), I ask in what part of its range will be the greatest angle of elevation?"‡ The sole answer is a small drawing of three curves, the greatest range being the curve

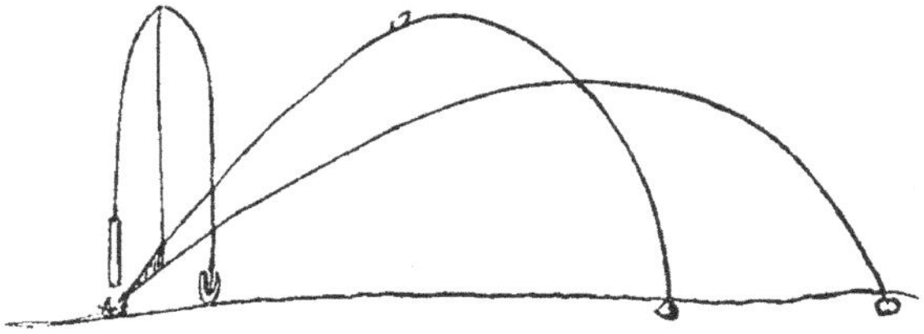

about midway between the perpendicular and the horizontal. Yet is this small drawing very remarkable when we come to examine it. In the

* Montucla, Hist. des Mathém.

† I, fol. 82.

‡ I, part ii. fol. 81.

first place, we see that he recognises the fact that the trajectory is a curve throughout its length; secondly, that a shot fired perpendicularly will not fall again on the spot whence it was fired. Simple as they may seem, these two propositions recognise the force of gravity, resistance of the air, and the rotatory motion of the earth.

Galileo is generally supposed to have been the first to point out that gravity causes bodies to move downwards with a velocity equally accelerated, and that when any body is projected forward, the same power forces it to describe a parabolic curve, unless so far as it would be diverted from that track by the resistance of the air. Copernicus is also looked upon as the first who, of modern times, revived the ancient doctrine of the rotatory motion of the earth. And Gassendi* took to himself the credit of applying to the motion of heavy bodies the theory of the motion of translation, a theory which made no progress until D'Alembert† showed that heavy bodies projected towards the zenith did not fall back again exactly where they had been projected from.

That Leonardo had taken all these matters into consideration, when making the sketch we have given above, is evident by the following remarks we find amongst his notes:—

* De Motu Impresso à Mutore Translato. Paris, 1642.

† Mémoires de l'Académie des Sciences, 1771 A.D.

"To define the descent or inequality of the intervals (of time in the fall) of bullets, I say, firstly, that through its gravity the descent of each bullet may be divided into similar divisions throughout the height (fallen). That in each division of this motion the bullet acquires an amount of velocity; whence such a proportion of gravital velocity makes a continued arithmetical proportion, because it proportions similarly the increase or decrease of velocity; whence I conclude that such spaces will be similar, because they always exceed or surpass one the other with similar increase, and for this (reason) water, which is poured from similar heights, does the same, also acquiring, as it does, similarly, at every instant of motion, an amount of velocity, whence, by arithmetical progression, it will go on increasing every instant of its descent, and for this reason it is that water, where it moves the most, the thinner it becomes. Letting fall water from a vase at a height of thirty yards, the course of the water will be thirty yards long, it is asked which will weigh the most, the first yard or the last." *

" *Of Resistance of the Air.*

"That atmosphere will become the denser which shall be struck by the greatest weight; although

* Libri, Hist. des Mathém. en Italie, vol. iii. note v.

b may be as great as *a*, nevertheless, if of itself doubly heavier, it will be a double weight on the

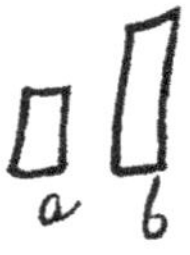

air which flies from it below; and the more dense it becomes below, the more rarefied it will become above, proportionally. (M, fol. 58.)

"*On the Motion of an Inuniform Body, which moves through the Air or through Water.*

"In a moving body of uniform matter, but of inuniform gravity, the heaviest part goes first. A pyramidal body, of a size uniformly diffformed, which shall be launched from a bow with the point in front, will immediately turn the base in the direction where its mass will move. (G, fol. 51.)

"Every heavy body will endeavour to fall by the shortest path (G).

"*On the Descent of Heavy Bodies combined with the Rotation of the Earth.*

"Let A be the body which descends through the elements, starting from A to go to the centre of the world, M. I say that this body, although falling in a spiral will not leave the straight line, with respect to the part from which it detached itself to go to the centre; for if the body leaves A to go to B, whilst going to B, and that it has been carried to C, the point A, whilst turning, has reached D; thus the body is still in the straight line which joins the point D to the centre of the world. If the body falls from C to F, the part D, which is that whence it began its movement, passes in the same time to E. Whilst the descent is made from F to G, the point E goes to H. Thus the body falls on to the earth always below the point from whence it started. It is a compound motion—it is at the same time straight and curved. It is straight because the body is always in the shortest line which extends from the point of departure to the centre of the elements. It is curved in itself; it is curved in each part of its path, and, consequently, it is curved throughout the whole of its path.

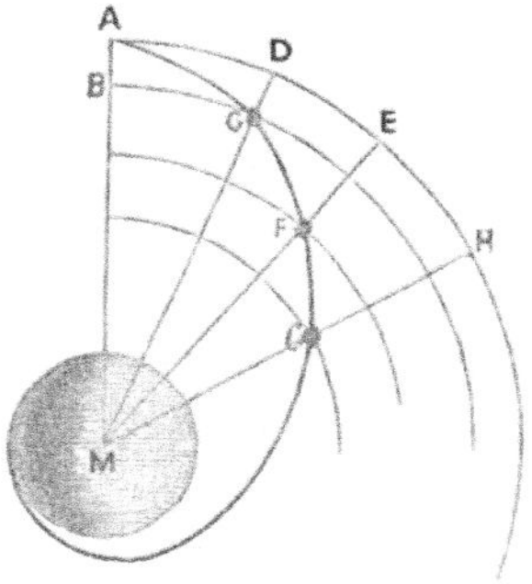

" That is why a stone which falls from the top of a tower does not strike the walls of the tower rather than the earth." (G, fol. 55.)*

These are certainly extraordinary observations to have emanated from the brains of the Tuscan painter, when we remember that even Galileo, more than a century later, argued that a body projected perpendicularly fell back again on the

* A translation of this has been given by Venturi.

same spot from which it was fired.* Whether or not Leonardo considered that the trajectory of a projectile was parabolical we have no evidence beyond his drawing; but as he ever looked first to practical results, it is probable that he thought but little of a theory which the ablest mathematicians, with the aid of all the resources of the most refined methods of analysis, have been unable to bring to more than an imperfect and distant approximation.† That, however, he considered 45° about the greatest angle is evident by the scale he made and the result he deduced from it:—

"If it is wished that the culverin bombard at *a*

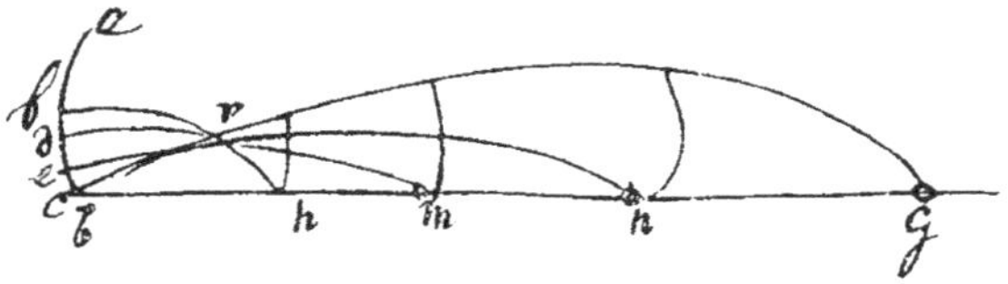

should throw its ball from *b* to *g*, I place the bombard in *c r*. I wish to fire to *n*, I raise the bombard in *d n*. To *m*, I will fire in *e r;* to *h*, in *f r;* for experience has proved the range of the bombard from yard to yard; and I made its scale divided into degrees, each (scale) of which will be 100, and each degree is divided into ten minutes, of ten to one, and each minute into ten seconds.

* Dialogue iv. prop. xiv. prob. vii.

† Hutton, Theory and Practice of Gunnery, tract 37.

Experience has shown in this way, following the scale, that the greatest range of the bombard will be shown when its ball falls by its own weight without force—that is, by its impetus, without rebounding." (I, fol. 81.)

And thus he argues on the more theoretical details of gunnery:—"The act of firing is made with divisible time, and as many more parts of time it lasts, so much the more powder burns, and the more the fire* crowds itself in such a machine, the greater will be the impetus and fury with which the ball will be thrown from it." (I, fol. 53.)

"A bombard which throws heavy stones is far more useful than a piece which throws leaden balls, for the stone causes the enemy much harm by its bounds and ricochets, whilst the leaden ball, by its weight, buries itself in the first spot it strikes, and, not rebounding, produces much less effect. (N, fol. 335.)† For as much as *a* rebounds, after

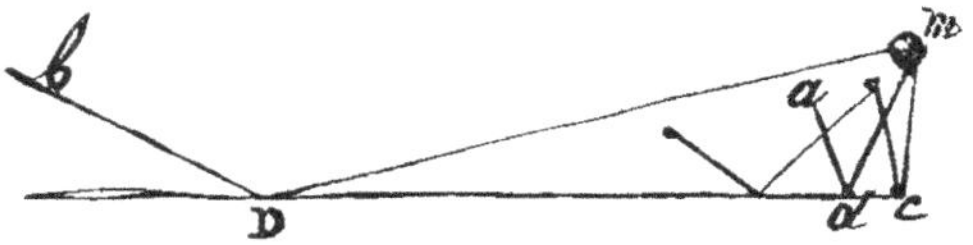

striking, more than *b*, so much further will *b* be

* There was then no idea of gases, yet not very long ago Count Rumford took credit to himself for this discovery.

† Venturi. Essai sur les Ouvrages Physico-Mathém. De Leonard da Vinci.

made to strike the ground from *c* than was *d* (G). Two equal bodies with equal velocities, I ask which will strike harder an object? The straight momentum *a b*, or the momentum

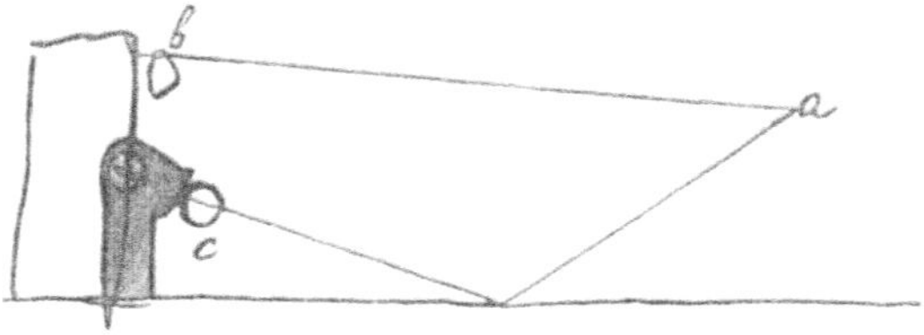

reflected against the slope *a c?* (I, fol. 22.) If stone or water is struck, the reflected motion of

the striking body follows in the direction which the striking body would have followed in consequence of its primary blow (K, fol. 2.) That part of the bronze condenses quickest in its mould which is the most liquid, and that is the most liquid which is the hottest, and the hottest is that which issues first from the furnace; therefore you should make, with the first pouring out, that portion of the *passavolonte* which is near the powder, sooner than that which is to form the muzzle (M, fol. 55.) Since, at the present day, artillery has increased three-fourths in strength, walls of fortresses should also have their resistance increased threefold. (N, 47.) For walls to resist the fire of bombards, counter-forts must be erected every ten yards, and the space between them should be filled with earth. The thickness of the counter-forts, as they extend outwards from the wall, should be diminished, to enable the earth to support itself, even when the artillery should have destroyed the wall in front (B, fol. 95, 96.) The farther the ravelin is (from the wall) the more it is exposed to cross-fire. The enemy, forming a lodgement in the trenches of the glacis which borders the ditch in L, B, H, K, they will erect

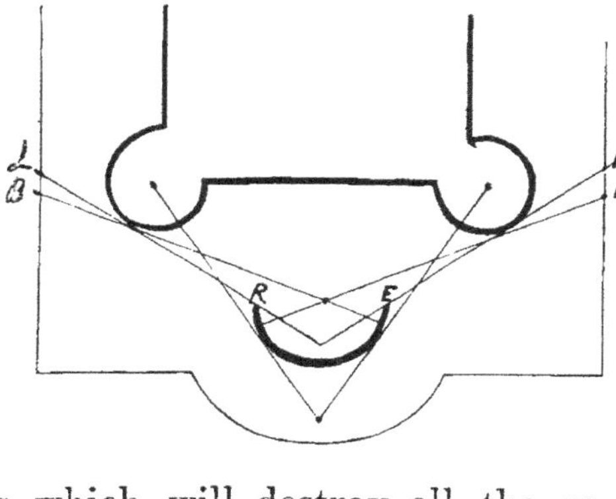

batteries which will destroy all the ravelin (N, fol. 42.)

"Do not place the mine* at the extremity of

* Gunpowder does not appear to have been used in connexion with the ancient art of mining until the middle of the fifteenth century. The earliest mention of it is in Pitti, B. Cronica. Firenze, 1720, 4to. p. 75, relative to an attack on Pisa in 1403.

the angle of a fort, because it would overthrow outwardly the part of the wall which forms the exterior angle. The expansive force, acting between two resistances, will overthrow, the exterior part, being the weakest; it will spare the interior part, which resists the more it is pressed." (N, fol. 128.) But, dissatisfied with improving on the engines of war of his age, we find the restless genius proposing the arch-thunderer. "This is a brass machine, which throws iron-shot with great noise and violence. It is thus used. The

third of this instrument consists in a great quantity of charcoal fire. When the water is very hot, you must tighten the screw over the vase *a*, *b*, *c*, where the water is; and by tightening the screw above, all the water will escape below, falling into the heated part of the instrument, and will immediately turn into steam so abundant and strong that it will be marvellous to see the fury of this smoke, and to hear the noise it produces. This machine throws a ball of the weight of a talent and six."* (B, fol. 33.)

* The works of Hero of Alexandria appear to have been a favourite study about this time. Of four MSS. contained in the British Museum one is assigned to the fifteenth and another to the sixteenth centuries. Doubtless Da Vinci had carefully perused them.

CHAPTER IV.

ALTHOUGH the art of fortification and of casting brass guns had made extraordinary progress by the middle of the sixteenth century, field artillery was still but little used. Armies did not like to

From the Wreck of the Mary Rose, sunk temp. Henry VIII.

encumber themselves with a *matériel*, which the almost general want of roads caused to be a decided obstruction to their movements. In A.D. 1522 we find the "Kinges ordonauns unable to pass over Stanes More towards Carlile;" * and the following year Henry VIII. raised the siege of Boulogne, because "His Highnes thinketh, that the wetenes of the cuntre, upon the rivers side, shall not suffre his army to march with artillery, either groce inough for batery, or sufficient for the feld; without which His Grace thinketh it were a great unsurtie to send theym thither, as they may

* State Papers, Henry VIII., vol. i. p. 105.

be percase constrayned to strike bataill with a more puissaunt hoost than is rekened on."* Consequently armies took but a small portion of field-guns with them; for instance, Francis I., when invading Italy in 1524, "had only vi canons serpentines; 3 colœuvrines bastardes; and 8 colœuvrines moyennes;" † and in A.D. 1569 the Prince of Orange, with 8000 infantry, and 14,000 cavalry, had only thirty guns.‡

Since the reign of Edward IV. England had ever been remarkable for its artillery,§ though in that country iron was still very expensive, and brass so scarce that, in 1530, its exportation was strictly forbidden, as it caused "great hurte of our soveraigne Lorde the Kyng in lackyng metall to make and furnish his ordynaunce." ‖ In 1543, however, Rafe Hoge and Peter Bawd succeeded in casting ¶ large iron guns at Buckstead, in Sussex,

* Ditto. More to Wolsey, p. 137.

† Arundel MSS., fol. 51.

‡ Letter found in the pocket of Condé at Jarnac. Capefigue's Réforme et la Ligne, tom. ii. p. 472.

§ Mémoires de Philip de Comines; Mémoires de Martin du Bellay, liv. i. A list of Q. Elizabeth's artillery is given in Egerton Papers, Camd. Soc.

‖ Statutes of the Realm, 21 Hen. VIII. c. 10.

¶ Hollingshed, Chronicles. Mushet, in his valuable work on "Iron and Steel," doubts whether cast-iron was known in England at this time; but there are some fine specimens of cast-iron guns time of Henry VIII. in the Repository at Woolwich. But the art was not generally understood even in the

an art they had probably imported from Flanders; and these mills soon became so celebrated, that, in 1587, the Earl of Leicester obtained by patent a three-years' monopoly for supplying the Low Countries with 320 tons of cast-iron ordnance, "to be made, gotten, and procured, in our countie of Sussex, or elsewhere in our dominions." *

But of all fire-arms the hand-gun underwent

Temp. Henry VIII.

the greatest improvement during the sixteenth century. The clumsy culverin, with its slow

seventeenth century, for cast-iron made at Dean was so brittle that the blow of a hammer would break it.—Bigland, Hist. Gloucestershire.

* Camden Soc. Egerton Papers, p. 122. In 44 Eliz., a very hot debate took place in the House of Commons respecting

match, had given way to the harquebuss, with its

From Gheyn, A.D. 1623.

pyrite wheel, the invention of a Nuremberg arti-

this exportation of iron ordnance. But as it was shown that Elizabeth made about 3000*l*. a-year by the patent, their valour cooled, and there it rested.—Hist. Collect. of the Four Last Parliaments of Queen Elizabeth, by Heywood Townshend. Lond. 1680.

ficer. The Spaniards, constructing these harquebusses of a heavy calibre, and using them with a rest, easily overthrew the lanzquenet, or man-at-arms, and spread the fame of their weapon throughout Europe. "It is my opinion, Sire," wrote M. de Forquevaux, French Ambassador at Madrid, to the King of Spain, on the 3rd of September, 1569, "that of the 4000 men, it would be good were 3000 to be harquebusiers; for war at the present day is carried on more by that than the pike, and the Spaniard makes a better use of it than any soldier in the world."* That the English did not make an equally good use of it arose from the fear which the Tudors had of the power bestowed on the masses by the free use of fire-arms, and to prevent which any paltry excuse was invented. Henry VII. had objected even to the cross-bow, because much damage was done by it to the King's deer.† Henry VIII. would allow no one to possess a hand-gun, unless he also possessed 100*l*. a-year, and then was stock and barrel together to be no more than twenty-seven inches long.‡ Edward VI. considered that "hayle shott" was pernicious, owing to much fowl and game having been destroyed.§ Singular to say, it was the wife

* Capefigue, Réforme et la Ligue, vol. ii. p. 469.

† Statutes of the Realm, 19 Hen. VII. c. 4.

‡ 25 Hen. VIII. c. 17; 33 Hen. VIII. c. 6.

§ 2 Ed. VI. c. 14.

of Philip of Spain who not only countenanced the men of England adopting the improved system of warfare, but even ordered every man worth 10*l.* a-year to furnish himself with a hand-gun ;* a practice by no means favoured by Queen Elizabeth, who issued, as late as 1596, a warrant for the Lord Keeper to pass commissions for the maintenance of archery.† But the day had gone by when the citizen could at any moment be driven from his counting-house to the muster parade. Some great emergency, such as the fear of the Spanish invasion, would still call forth all his energy, and induce him to cheerfully put on helm and breastplate, but he preferred the soldier doing the soldier's work, and all Elizabeth's proclamation could not rescue the artillery-ground from the ridicule of the London apprentice.‡ Yet the moral strength which a nation acquired by the free use of fire-arms was fully admitted about this time by Charles IX. of France, when, in 1571, he issued an edict

* 4 Mary, c. 2 & 3.

† Egerton Papers, p. 218.

‡ "*Stephen.*—I protest, as I am a gentleman, but no soldier, indeed——

Well-bred.—No, Master Stephen, as I remember your name is entered in the Artillery Garden.

Stephen.—I, Sir, that 's true. Cousin, may I swear, as I am a soldier, by that?

E. Kno'well.—O yes, that you may ; it's all you have for your money."—*Every Man in his Humour*, Act iii. Scene 5.

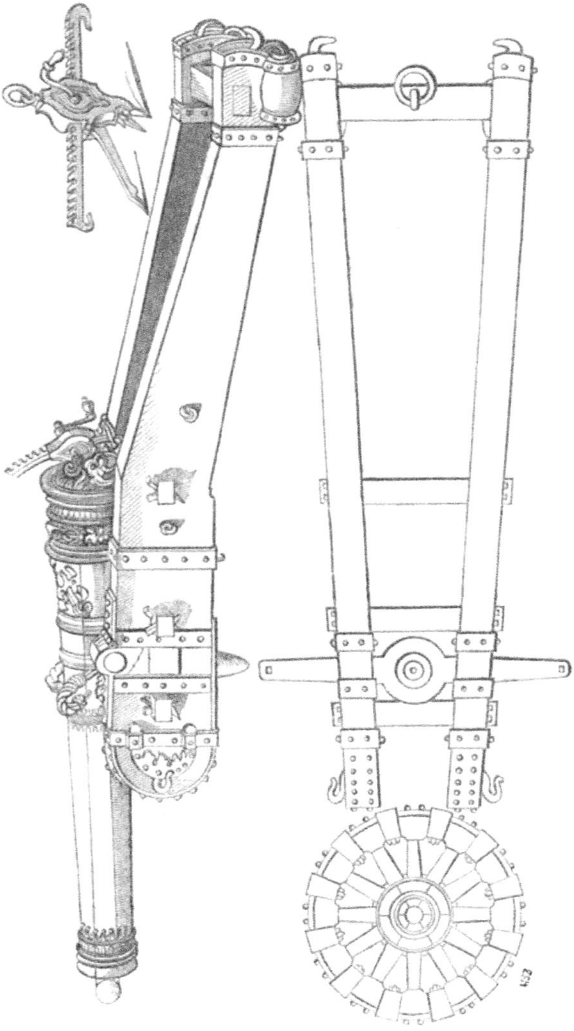

Field Gun, Sixteenth Century. Albert Durer.

prohibiting every person, of whatsoever position

or condition they might be, from making or casting any piece of artillery, without patent letters and leave obtained from the "Master of Artillery," with the exception of his city of Paris, whom he did not dare include in the order.* But such edicts are at all times of little use; and in France, during the civil wars of this period, Protestants, Royalists, and Leaguers, each in their turn, boasted of their artillery, according as the fortune of war placed towns of importance in their possession; and the Protestants placed especial reliance on their harquebusiers, whom they endeavoured to make as numerous as possible. In England this view of the matter was but little thought of, until the civil wars forced the minds of the people towards it; and one of the first acts of the Long Parliament, 1641, was "for the making saltpetre and gunpowder within the kingdom, which was then a part of the royal prerogative;† and not only considerable, as it restrained that precious and dangerous commodity from vulgar hands; but as, in truth, it brought a considerable revenue to the crown; and more to

* Capefigue, Réforme et la Ligue, tom. iii. p. 171.

† In the debate respecting exportation of ordnance in 44 Eliz., referred to before, it was a question whether it was or not part of the royal prerogative. The lawyers decidedly said it was not, referring to 33 Hen. VIII. c. 7, and 2 Ed. VI. c. 36, to support their view.—Hist. Collect., by Heywood Townshend. Lond. 1680. The Stuarts appear, however, to have made it so.

those whom the crown gratified and obliged by that licence. The pretence for this exemption was, the unjustifiable proceedings of those (or of inferior persons qualified by them) who had been trusted in that employment; by whom, it cannot be denied, many men suffered; but the true reason was, that thereby they might be sure to have in readiness a good stock in that commodity, against the time their occasions should call upon them."*

From this feeling respecting the value of artillery in warfare also arose the struggle between the King and Commons respecting Hull and the Tower of London. The magazine at Hull contained all the arms and ammunition belonging to the army, which had been lately raised against the Scots. The Commons were exceedingly anxious to have them removed to London, as Sir John Coniers, Lieutenant of the Tower, was entirely of their party; but the King objected, and for fear of accidents he appointed the Earl of Newcastle Governor of Hull; but that nobleman was forestalled by the Commons, who had dispatched Sir John Hotham, and on the King's raising his forces at York, they ordered Sir John to deliver the stores to the fleet under the Earl of Warwick, who brought them to London.†

The loss of Hull was a serious matter for the

* Clarendon, Hist. of the Rebellion, book iii.

† Ibid. books iv. and v.

royal cause, but even then, had Charles had the firmness to have kept the Tower in his own hands, he might have saved his throne. It was by no means a fresh feature in the annals of this country, that the King and his subjects should dispute respecting the guardianship of that fortress. From time immemorial its massive walls had been considered but as a bridle upon the City of London, and the quarrels between the City authorities and the Lieutenant of the Tower had been numerous. Sometimes the citizens would summon the Lieutenant for over-exacting dues; sometimes the Lieutenant would enter the City and forcibly quell a riot; sometimes the City would be placed in command of the Tower; sometimes the City under the Lieutenant of the Tower. At other times would it be debated whether the Tower was within the jurisdiction of the City or not.* These constant feuds had therefore, in the course of time, given an importance to that royal fortress which gave great moral preponderance to its possessor, and became paramount when it was the only arsenal of the kingdom. Charles I. made an attempt to retain it, by appointing first Colonel Lunsford, and afterwards Sir John Byron, as Lieutenant. He could not have made a better choice than the latter, for the bluff cavalier swore "that

* Antiq. Legib. p. 156; Hall, Chronicle, A.D. 1517; Maitland, Hist. of London.

he had taken an oath to maintain the privileges of the Tower, 'which extend over Tower Hill and over all the water that is against the Tower,' and therefore he will suffer no guards there without the King's warrant. Besides, he knew no power that Parliament had in the Tower, and he would obey no command but the King's."* But the King gave way, Byron was succeeded by Balfour, and when Charles I. raised his standard at Nottingham he had not with him one barrel of powder or one musket.† If, however, the people had begun to value portable fire-arms, they did not as yet understand how to use them, and the first effect of this universal employment of the hand-gun was to overthrow all and every system of warfare.

Mr. Darwin, in his "Natural History of the Voyage of the Beecher," gives an amusing account of the effect of fire-arms on the aborigines of Terra-del-Fuego. They could not be made to understand how that, standing a long way off, they should hear a noise and see a little smoke, and then that they should be mortally wounded. It was to them quite incomprehensible, and they therefore felt no alarm at the sight of a musket. But once they became convinced of the danger, their terror was proportionate. This feeling was

* Verney Papers, Camden Society, p. 152.

† Clarendon, Hist. of the Rebellion, book v.

curiously exemplified in the total demoralisation of all armies, from the middle of the sixteenth century till near the end of the seventeenth. The Swiss complained that it was not their nature to stand patiently under the fire of artillery, but to march straight onwards.* Montluc's soldiers would say to him, "Let us engage, sir; it is better to die sword in hand than to be killed by artillery;"† whilst the Italians would call it that *stromento d' inferno*;‡ and this nervousness increased still more when they saw their ablest knights, such as a Bayard, a Guise, and a Sydney, fall by an unknown hand. Tacticians wrote elaborate treatises on forming squares and stars of piquemen and harquebusiers; but human nature was too strong, and there arose that disorderly system of skirmishing, when each man looked but after his own safety, which he thought must lie in his own carbine and petronel, and consequently loaded himself with quick-match and powder-flasks; a system of warfare well described by Elliot Warburton, in his "History of Prince Rupert," as "limited to the wasting of gunpowder, and in making a noise,"§ and the amount of quick-match

* Mémoires de Du Bellay, liv. x.

† Montluc, liv. ii.

‡ L'Archibugio è lo Stromento d' Inferno, se è una bocca di fuoco. Nuovo dizionario. Geneva, 1676.

§ At Nazeby the Royalists had twelve guns, and these they

used was certainly enormous, considering the damage done. After the battle of Lansdowne, when Waller laid siege to Devizes, complaint was made that there was only 150 cwt. of quick-match in the town, whereupon "diligent officers were directed to search every house, and to take all the bed-cords they could find, and to cause them speedily to be beaten and boiled;" * a deficiency not to be wondered at, when their match was kept constantly burning, a mistake the Roundheads felt the penalty of when its glimmering caused them to be defeated by Sir John Byron at Burford.†

It required a Gustavus Adolphus and a Cromwell to overcome this lax state of discipline, and once more to teach the soldier how to face death in the ranks; but then, the one with his Lutheran Swedes, the other with his psalm-singing Ironsides, had a wonderful material to work upon, for even at the present day the Southern gets impatient at being kept any length of time under fire. The success of these two great leaders forcibly impressed on every European government the total inutility of

did not even bring into the field. One was a 24-pounder, drawn by twenty-six horses. So little, indeed, was thought of it, that in the plan of the battle, drawn out by the King's command, no position is told off for the artillery.—Warburton's Prince Rupert, vol. iii. pp. 104 and 111.

* Clarendon, Hist. of the Rebellion.

† Warburton's Prince Rupert, vol. ii. p. 91.

a nation relying upon its feudal levies for defence. The English nobleman, with his tenantry, had been cut down by the Red Coats, and the French *gentilhomme* had learned to dread the mercenary *Reître.* As to a city being protected by its wall and turreted gateway, and the citizen keeping watch and ward, it had become simply ridiculous.* The men who could understand Shakespeare and Ben Jonson would not be bound by the laws and regulations affecting those who could witness a mystery with pleasure. They wanted handsomer and larger dwellings, and the liberty they sought was not whether their cattle should be allowed to graze in the city ditch, but that the road to distinction and affluence should be open to them as well as the noble. The city wall was an impediment in the road to freedom, and it was allowed to crumble quietly away, except in such towns as from their nature and position rendered their being strengthened according to the new art a national necessity.

* Discours des Villes, &c., assaillies et prises par la force de l'Artillerie sous Henri II. et Charles IX. Paris, 1568. No. 32,148 of the Bibliothèque, Hist. de la France.

CHAPTER V.

WARFARE was about to assume a new feature. Fortresses had been so strengthened against artillery of that day, that generals often preferred leaving them alone; and, if it was absolutely necessary to

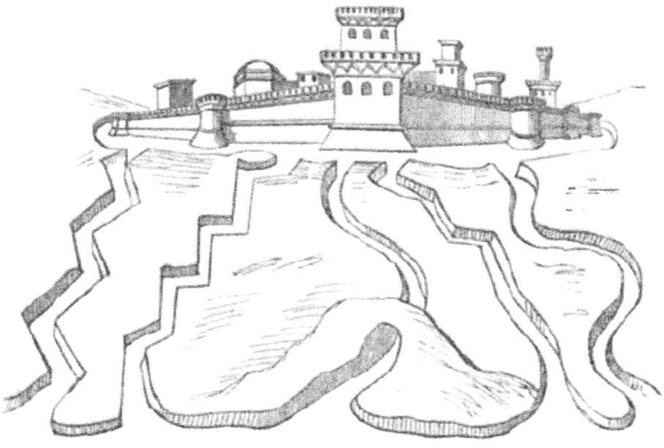

Siege Operations, Fifteenth Century. Giorgio Martini.

besiege one, the operation was most tedious, and more lives were lost by disease than the enemy's fire. It had, therefore, become necessary for the engineer to bring offence on a par with defence, and Sebastien Leprestre de Vauban, the successful discoverer, became the greatest man of his age.

This military genius appears to have been endowed with a peculiar aptitude for destruction. He would sit down before a town, and, at a glance, seize upon its weakest point, calculate to a nicety the strength of its walls, and whilst he was inquiring into the effect of a pickaxe in that particular soil, he would trace the plans of enormous siege-works by which to protect the besiegers in their approach; for, like a great general, his first care was the preservation of his men, upon whom he was fully aware rested his only chance of success. It was with a view to this that, at Maestricht in 1673, he introduced his famous parallels and places of arms, and that at every fresh siege he went on adding to his new system, sometimes a cavalier in the trenches, at others a new mode of sapping or demi-sapping, and, finally, he astonished the world by an ingenious application of the force of gunpowder—ricochet firing.

The increased resistance of masonry had at an early period induced artillerymen to obtain a flank fire, and we also find Albert Durer and Leonardo da Vinci had studied the advantages resulting from the ricochet of shot. But Vauban was the first to unite the two, so as to create a formidable system of attack. Throwing up his batteries on the flank of a bastion, he lessened the charge and elevated the guns so that the shot went just clear of the enemy's parapet and dropped within their works,

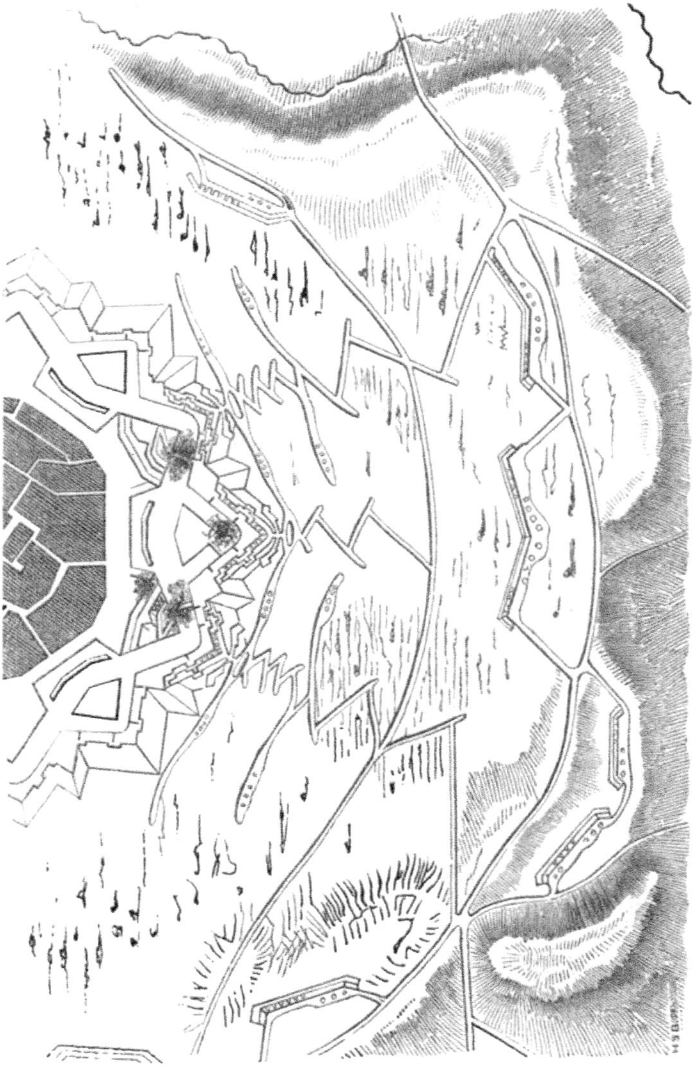

Vauban's system of Attack.

when, bounding along, they raked the whole length

of a face and disabled its guns. The effect of this simple idea was to cause perfect consternation, not only in the minds of engineers, but also in that of governments. The former were vexed that their numerous self-supporting fronts, well strengthened with cannon, should be rendered useless by shot hopping along and destroying their armaments, and they proposed various ineffectual remedies; amongst others earthern mounds at certain intervals, called "traverses," but shells were then used and the traverses destroyed. Further than this, however, the evil could not be remedied, for all these suggestions cost money. In the preceding century the erection of the citadel of Antwerp had cost upwards of 1,400,000 florins, and the fortifications of such towns as Malines or Orleans averaged about 5,000,000 florins.* Since then the various improvements introduced ran up a front of fortification of four hundred yards in length to some 40,000*l.*, and the alterations proposed by Vauban to counteract his own plan of attack doubled that sum. Yet was he economical compared to many who attempted to improve on his alterations. It is not my intention to investigate the numerous theories brought forward by men skilled in military science for this purpose, nevertheless I shall take

* Discours Politique, &c., du Seigneur de La Noue. Basle, 1587. A Plan of Antwerp found at Simancas is published in Gachart's Doc. inéd. de la Belgique.

two of their number, Montalembert and Carnot, in order to exemplify the ideas of the most able amongst them. Montalembert was a decided opponent of the bastion system, and proposed to substitute several *enceintes* situated within the main ditch, and a formidable artillery placed in bomb-proof casemates, the latter serving likewise as a protection for the garrison, for stores of all kinds, and even as a refuge for the inhabitants. Carnot, on the other hand, proposed an alteration in the trace or outline of the polygon, the suppression of the interior revetment of the covered way known as the counterscarp, to detach the scarp wall from the rampart, and to make the latter without a revetment, and cause destructive personal conflict with the besiegers by means of frequent sorties. There was nothing very new in Montalembert's theory, and its expense alone, some 300,000*l.* per front, put it out of the question of practical proof. As to Carnot's, its chief advantages were the detached wall and destructive personal conflict. The latter is a dangerous kind of warfare, and requires an army; the former was overcome by a little science. Carnot had fancied that by detaching the scarp wall from the rampart, that when attempting to surprise by escalade or by main force, the assailants having not only to mount a wall twenty-four feet high, but also to lower themselves down a similar depth, would suffer greatly from the musketry-fire which could be directed through its loopholes; and many fortresses were altered to this

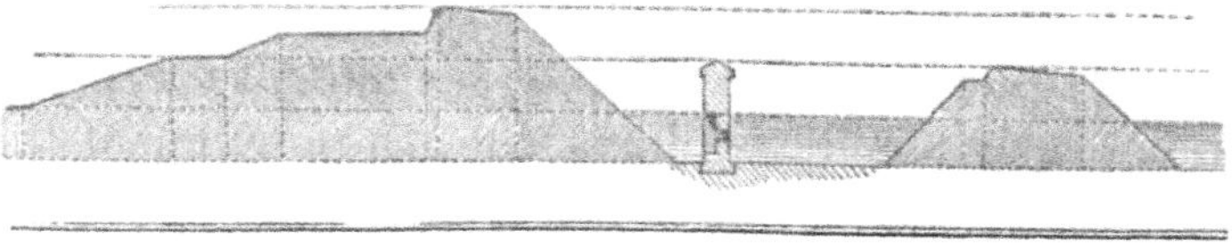

Carnot's System.

system at a very great expense. One important fact had, however, been forgotten, which was that this wall might be destroyed previous to the assault. On the representation of that distinguished officer, Sir Howard Douglas, experiments were carried on at Woolwich in 1823, in order to determine whether or not it was possible to breach this wall by the fire of heavy ordnance directed over the covering-work. A Carnot wall was built, and then siege batteries were erected against it, consisting of eight 68-pounder carronades, three 8-inch and three 10-inch howitzers, at distances of 400 and 500 yards. In six hours after the opening of the fire, a practicable breach fourteen feet wide was effected.*

But to explain this we must revert to the progress of artillery during the eighteenth century.

The minor details of field-artillery had under-

* Douglas on Naval Gunnery, edit. 1851, p. 68. The full details of this important experiment are to be found in a Report on the Practice against Carnot's Wall, &c., in August 24. Lithographed at Chatham.

gone great improvements under the able generals who commanded during the Thirty Years' War; and gunpowder was made so much stronger, that the 24-pounder gun of the seventeenth century was nearly as efficacious as the 48-pounder of the sixteenth; but the scientific application of gunnery was paid little attention to. Robins, who had carefully studied all the writers on the subject of the seventeenth century, who were exceedingly numerous, only found four out of the number who had actually tried the ranges of shot and shells at various elevations, and only one, Eldred, Master-gunner of Dover Castle, who came any way near to truth;* the mass of them, although Galileo had shown that a projectile would, in its flight, move in the curve of a parabola, unless forced to deviate from that track by the resistance of the air, persisting still in stating that the air's resistance was *nil;* and amongst this class of theorists we find even Halley and Blondel.† Yet had the difficulties attending such experiments been considerably lessened by Newton and Descartes. The former, in his "Philosophiæ Naturalis Principia," had clearly demonstrated the law and quantity of the resistance of the air to slow motions, and how much greater it would proportionally be with swift. Descartes had applied algebra to the theory of curved lines, which enabled the mathematician

* Robins, Mathem. Tracts. Preface. † Ditto.

to calculate them in an easy manner. But to apply the theories of master-minds requires a master-mind; and a Vauban had to point out not only the advantages of ricochet, but also how to breach systematically, by first sapping the whole wall as low as possible before directing the fire to the upper portion.* As if overcome with the practical results of these two ideas, artillerymen and engineers once more relaxed their exertions, until a Robins published in 1742 his small treatise, entitled "New Principles of Gunnery," and pointed out strange unthought-of facts, which, translated into German and French, with commentaries by Euler and others, excited the greatest attention on the continent, and gave rise to an amusing controversy amongst French artillerymen, which is not without its moral at the present day.

In 1732 the French artillery had been organised upon a system admirably suited to the warfare of that period by Jean Florent de Valière.† But, during the Seven Years' War, many of its officers had sought service in the armies of Maria Theresa, and amongst these was Gribeauval,‡

* Expériences de Bapaume, p. 5.

† There were two distinguished officers of the name—Jean Florent, who became director-general of artillery in 1720, and died in 1759, and his son, Joseph Florent, who was director-general in 1755.

‡ Jean Baptiste Vaquette de Gribeauval became Lieutenant-General Inspector-in-Chief of the French Artillery in 1776.

who was entrusted with the command of the artillery under the Prince of Lichtenstein. There he became acquainted with the system of light field guns, which Austria had been forced to bring into the field to keep up with the rapid movements necessitated by the manœuvres of the King of Prussia. On his return to France, after the war, Gribeauval strongly impressed upon Choiseul, then Minister-at-War, the advantages which would result by following in the same footsteps,—a change which had already been solicited by several general officers who had served in the late wars. Choiseul, who was already biased in its favour, ordered a series of experiments to be carried on at Strasburg; but in order that the subject might be examined with impartiality, these were to be as public as possible; and the only general officer of artillery appointed on the committee was Du Mouy, one of the most determined partisans of the system of 1732. The principal point to be determined was, how far guns could be lightened so as to place the French artillery on a par with that of the other nations of Europe as to its manœuvring, and yet retain that superiority in its power which they had so long boasted of. The first act of the committee was to place the 16-pounder on one side as a siege-gun, and thus reduce the number of field guns to 4, 8, and 12-pounders; for, from the experience acquired during the late wars, it had been found that owing to the short effective range of musketry, 600 yards was sufficient for any artillery purpose in the field, and, consequently, that 1000 yards would be ample. As, therefore, the only advantage resulting from the Valière system was a range greater than this, and which sometimes proved detrimental owing to the inducement to begin firing at too great a distance, it was considered that guns might be shortened considerably.* Gribeauval, basing his calculations on the effect of Austrian and Prussian field-pieces, the former 16 and the latter 14 calibres in length, with respective proportions of 120 and 100lbs. of metal to each 1lb. of shot, proposed that the French should be 18 calibres in length, with 150lbs. of metal to 1lb. of shot. Two guns of each calibre were tried, and with such success that, in 1765, 150 guns were ordered to be cast on the new model. Nevertheless, in spite of two ordonnances of the king, of the verdict given by a board of marshals, consisting of men eminent in their profession, such as Rohan-Soubise, Richelieu, De Broglie, and De Contades, not only the regiment of artillery, but even the Académie des Sciences was rent by the partisans of the old and new systems. Voluminous works were published, pamphlet succeeded pamphlet, and the Encyclo-

* Mémoire de M. Gribeauval qu'il a presenté à M.M. les Maréchaux de France.

pædists were alternately canvassed by either party.* The system of 1732, said the partisans of the old school, had supported the pre-eminence of the French artillery for thirty years, more especially at Fontenoy, where the formidable English columns, almost masters of the field of battle, had been arrested in their forward movement by its long range; besides, was it not far preferable to take into the field a few well-served guns, possessing long range and great accuracy in their fire, than a multitude of light pieces, which could but encumber the army on the march, and impede its movements by the scarcity of provisions it might entail? On the other side it was contended, that for the French artillery to keep up that eminence which was so boasted of, it was necessary to keep pace with the style of warfare then in vogue; that the tactics introduced by Frederic of Prussia necessitated greater mobility in the material of an army, whilst few pieces weighing 33 and 22 cwts.,—which were the respective weights of the old 8-pounder and 4-pounder, placed on a two-wheeled carriage, could keep pace even with infantry,

* Mémoire lu à la séance de l'Académie des Sciences, 13 Novembre, 1751; Observations faites par M. de St. Auban, Lieut-General de l'Artillerie, sûr ce Mémoire, &c.; Encyclopedie, Art. Cannon; Mémoire touchant la Supériorité des Pièces d'Artillerie longues et solides, &c., par Valière, Lieut.-General, lu à la Séance de l'Académie des Sciences, 19 Août, 1775, &c.

the manœuvres of which had undergone such important changes; that the improvements in muskets and the greater reliance placed on infantry-fire, having brought about the extended formation of three deep, a greater number of guns were required to neutralise its effect; that though the system adopted in 1732 possessed long range and accuracy, it was still possible to reduce their length and yet maintain these points,—facts pointed out by Belidor, and clearly demonstrated by Robins. Nevertheless, the old school maintained their points so obstinately, and poured forth such a torrent of battles and sieges, in which the heavy artillery had succeeded wonderfully, that one of their opponents wittily compared them to M. de Turenne's horse, who, after having been in many an action, was none the wiser;* and Gribeauval completely refuted the arguments brought forward respecting the probable decrease in the range of the shorter pieces, by putting into practice Robins' theory of windage, which he reduced from two-twelfths to one-twelfth; and he subsequently obtained leave to reduce the weight of the gun about one-fourth. Other nations were not idle spectators of these discussions. Experiments were carried on at Turin by the Chevalier d'Antoni, and in England, in 1775, by a committee,

* Lettres d'un Officier d'Artillerie à un Officier Général, par M. Troncon de Coudray. He was subsequently brigade-major in the French Expedition to the United States, where he was drowned.

amongst whom we find the names of Dr. Hutton, Major Bloomfield, afterwards General Sir Thomas Bloomfield, and Captain Congreve, better known as the first Sir William Congreve; but these did little more than verify the truths of Robins' theories respecting the resistance offered to the flight of projectiles by the atmosphere, and the variations caused in the direction of that flight by the action of the powder on the shot whilst within the bore of the piece, combined with the air's resistance.

The improvements in siege ordnance were, if of slower growth, yet of greater importance. The inefficiency of iron in withstanding heavy charges had caused 42-pounders, then the largest guns in use, to be still principally cast of brass; but their weight, some 64 cwt., rendered them, of course, very expensive,* and so difficult to convey as siege-train, that seldom was any gun larger than a 24-pounder made use of on such occasions. But the extraordinary progress of the iron trade of this country in the latter half of the eighteenth century not only enabled it to add considerably to the number of our heavy ordnance, but to greatly increase their effect. To enter into details on this subject would be to write the history of the iron trade between 1750 to 1800;† suffice it to say, that the introduction of pit-coal instead of charcoal,

* Appendix E.

† Scrivener, Hist. Iron Trade; Mushett, Papers on Iron and Steel.

and Watt's double-power engines, instead of water-wheels working forcing-pumps, assisted in producing quantities of iron hitherto undreamt of, that thousands of practical minds, attracted by the great competition which arose in consequence, turned their attention towards improving the quality of their iron, so as to obtain higher prices in the market; that by these united energies such valuable improvements were made that, in 1785, it was found necessary to enact that any one enticing abroad a workman who had wrought in the iron or steel manufactures became liable to a penalty of 500*l.*;* that these improvements in the quality of iron enabled the Ordnance to do away gradually with brass heavy ordnance and substitute iron, and, about 1780, to introduce short light iron guns, called "carronades," from the foundry they

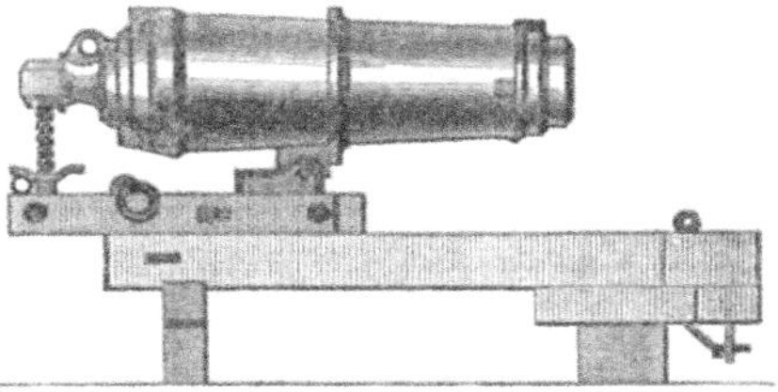

were made at, and "smashers," from the service they rendered when engaged at close quarters, by firing 68-pound shot out of a gun weighing only 36 cwt.; and that, at the beginning of the present century, small vessels were thus enabled to carry large ordnance, and siege-trains to transport guns of

* 25 Geo. III. c. 6 & 7.

sufficient power to destroy the fortifications which had been erected to withstand the old 24-pounder.*

These were great results, yet were they shortly to be incomparably surpassed.

* As to the services rendered by carronades on their first introduction, see Sir Howard Douglas on Naval Gunnery, edit. 1851, pp. 68 and 150.

CHAPTER VI.

Robins, amongst other subjects, had turned his attention to the nature of the rifled pieces then used in many parts of Europe, and the advantages which they presented in obviating to a great extent the deflection of projectiles from their first direction; and although during his time rifles were but rude weapons, the number and depth of grooves differing according to the fancy of the workmen, and that the bullet, slightly larger than the bore, was forced down by a strong rammer and a mallet, nevertheless he concludes his paper by "predicting, that whatever state shall thoroughly comprehend the nature and advantages of rifled barrel, and having facilitated and completed their construction, shall introduce into their armies their general use, with a dexterity in the management of them; they will by this means acquire a superiority which will almost equal anything that has been done at any time by the particular excellence of any one kind of arms; and will, perhaps, fall but little short of the wonderful effects which histories relate to have been produced formerly by the first invention of fire-arms." *

The first country to take up the theory of rifles was France, and since the peace of 1815 experiments on an extensive scale have been carried on in that country, and the careful studies of a few energetic officers have thrown a light on the subject, exceeding anything thought of even by Robins. Delvigne first pointed out the idea of causing bullets to force themselves into the grooves by means of the expansion of the powder acting on a hollow at the base of the bullet; but difficulties arising in the endeavour to find out what form should be given to this hollow, so as to secure the certainty of expansion without the bullet being distorted by the force of the gunpowder, M. Minié placed in the hollow a spherical iron cup, which produced the required effect; and M. Tamisier, considering bullets often deviated

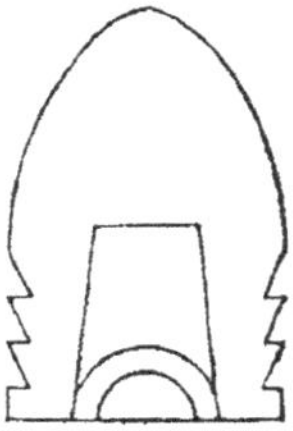

from the correct line of flight, on account of the gravity of the bullet being affected by air-bubbles, &c. in the lead, after a great number of experiments, proposed certain circular rings round the base of the bullet, intended to have the same effect as the feathers of an arrow, which proved successful. Series

* Robins, Mathem. Tracts.

of trials had also been made as to the length, number, and curve of the grooves, the charge of powder, &c.; and the results of all these trials combined produced the rifle of which so much was spoken of in 1851. It was still, however, far from perfect; and though the Marquis of Anglesea, when Master-General, ordered 28,000 of these arms for the British army, it was only intended as a temporary measure. Lord Anglesea went out of office in Feb. 1852, and was succeeded by Lord Hardinge. This able officer followed in his predecessor's footsteps, and requested all the leading English gunmakers to send in specimens of the arms they would propose for the use of the soldier, its weight being limited to 9lbs. These were submitted to a committee at Enfield Lock, consisting of Colonel the Honourable Alexander Gordon, Captain Warlow, R.A., and Mr. James Gunner, inspector of small arms, than whom no one is more thoroughly master of the subject. No single weapon presented the required perfection, but from a general modification resulted the beautiful Enfield rifle in use in the British service; and the committee further adopted as the service bullet one proposed by Mr. Pritchett, junior, the London gunmaker, which had its outer surface perfectly plain,

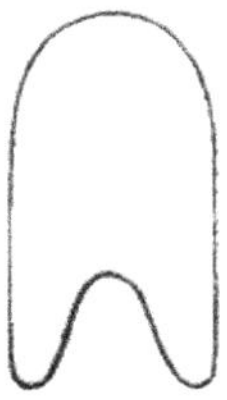

and a hollow so constructed as to do away with

the necessity of the iron cup.* But before such a musket could be made, it must be borne in mind that many other points had to be thought of besides the theory of projectiles. Cort had to take out in 1783 his patent for puddling and rolling of iron, and many an iron-master had to improve on Cort's patent before Mr. Marshal of Wedgebury could supply that splendid iron which enables our barrels to be so light though so tough. Sheffield had to devote its energies to steel before we could get our present locks and bayonets; Germany and Italy had to be ransacked for sound walnut timber; lead had to be purified, and compression bullet-machines perfected by Anderson, before the cannelures or iron cups could be done away with: in fact, the small arm of the present day is the result of the progress of science in all its branches. And so is the present heavy ordnance.

The French had always found great difficulty in making iron ordnance.† Montalembert and

* Remarks on Rifles and Rifle Practice, &c., by Col. Hon. A. Gordon. The Rifle-Musket. Chapman and Hall. Since then, for some reason, a wooden plug has taken the place of the Minié cup; but why or wherefore I could never understand.

† Recherches sur les Bouches à Feu en Fonte de Fer, par le General Tirlet; Thierry, Application du Fer aux Constructions de l'Artillerie; Paixhains, Constitution Militaire de la France, Note F.

Buffon had endeavoured to bring them into use for the land-service, but without success. Indeed, they were even accused of wishing to make their fortunes by their iron-foundries at the expense of the lives of artillerymen; and in 1849, out of 16,495 guns which the French possessed, 13,770 were of bronze, in importing which, from 1840 to 1845, they spent 6,616,300 francs.* This use of brass, at a time when that of ordnance had extended itself over such a multitude of fortified points, makes it appear strange that a 24-pounder, which cost 7250 francs, should be employed when a cast-iron one could be got for 1500 francs. The fact is, that during the long wars of the Revolution and of the Empire the manufacturing powers of France had got annihilated, and their workmen were fighting on the Po or the Vistula, instead of endeavouring to improve their iron ware. Guns of a moderate weight could not, therefore, be made sufficiently strong to resist with safety the necessary charges of powder; and whilst, in 1812, at Badajoz, British 24-pounders fired 129 rounds in 104 hours, in 1835 it was dangerous to fire a French iron gun 100 times in twenty-four hours.† In this dilemma it occurred to General

* In England, from 1828 to 1848, only 40,493*l.* was expended in brass ordnance, at 140*l.* per ton, compared to 320,688*l.* of iron, at 15*l.* per ton.—Parliam. Commit. on Ordnance Estimates, 1848, pp. 971, 974.

† Observations on Heavy Artillery, by Lieut.-Colonel Sir

Paixhains, a distinguished artillery officer, that if French guns could not stand the heavy charges then in use, that he might, perhaps, obtain all the necessary results with less charges. This fact had to a certain extent been demonstrated with the English carronades; but then they were useless except at short ranges, and Paixhains was desirous of even extending his range. Taking into account the fact that all the attention of artillerymen had been directed towards reducing the mass of a projectile and increasing its velocity, he thought he would obtain the same momentum by increasing the mass and decreasing the velocity. For to increase the velocity of a projectile the charge must be increased, and consequently the resistance of the metal must be increased, thereby adding to

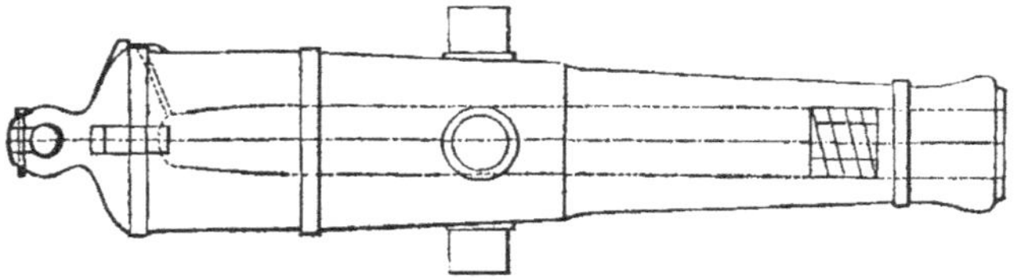

its weight and difficulty of transport. General Paixhains proposed to fire an eight-inch hollow shot with a 32-pounder charge, and consequently out of a gun weighing no more than a 32-pounder.

John May, K.C.B. London, 1809; Application du Fer aux Constructions de l'Artillerie, par A. Thierry, Capitaine d'Artillerie. Paris, 1834.

This was tried at Brest in 1823. The first round fired into a line-of-battle ship made a breach of 150 square feet, the second carried away the main-mast and iron-work, weighing 130lbs.* Now, General Paixhains thought he had made a great discovery when the wonders of the new rifle showed him there was much more to be done, and he therefore heads a note to one of his works, "*Sur un nouveau fusil qui est fait et sur un nouveau canon qui est à faire.*"

But then Paixhains was an able man, for there were numbers of petty minds who insisted the acmé of fire-arms had been attained, and beheld visions of field batteries in an action disabled by the total loss of their gunners, picked off one by one at 1000 yards. War was to be for the future all skirmishing, and men were to stand in the most amiable way to be picked off. But the world and all within it keeps ever progressing, and men still go on thinking, and experimenting, and discovering. Amongst others, Lancaster, the London gunmaker, has made a bold attempt to be the successful discoverer of the gun that is to be. Thinking over the theory of the rifle, a most ingenious thought occurred to him. A barrel was rifled in order to give a twist to the bullet, then why not form the

* Constitution Militaire de la France, Paixhains, Note H; Esprit des Institutions Militaires; Marmont. p. 88.

barrel itself so as to give this twist? and to elucidate this principle he introduced his elliptical-bored carbines. They have proved highly successful, and the Lancaster rifle shoots as well as the Enfield; and in one respect, to which we shall allude presently, it is even better adapted as a soldier's weapon. Lancaster next reasoned that the principle which answered so well with the carbine would do equally well with large ordnance; but here he had to encounter the serious impediment presented by the size of the projectile—a difficulty which a great many men before him had endeavoured, though unsuccessfully, to overcome. For that which renders the rifle so valuable is the ease with which lead adapts itself to the form of the harder metal, of which the barrel is composed, but it was impossible to think of eight-inch lead shot; and iron cased in lead had been more than once rejected by ordnance committees, as complicated and expensive. Lancaster, undismayed by the obstacles to be overcome, endeavoured to subdue iron, so as to make conical shells, 75lbs. in weight, sufficiently tenacious to resist the shock of explosion; that is to say, not to smash from the friction against the bore, when taking the rotatory motion. In August 1850 experiments were carried on at Shoeburyness, to test the results of his labours. Of seven shells fired, six burst to pieces on leaving the gun,* and the seventh only wabbled some 800 yards. Nothing daunted, Lancaster set to work again, and in December 1852 he succeeded in obtaining a range of 5600 yards, first graze.† There was still, however, great difficulty in welding the different parts of these wrought-iron shells, so as form one perfect whole; and as late as 1854 their cost, some 20*l.*,‡ prevented their being used to any considerable extent; but since then experience has not only shown how they can be welded of one piece, but has diminished their cost to one-twentieth; and the Lancaster gun is as yet the most formidable piece of heavy ordnance possessed by this country. There appears also to be no reason why a principle which has succeeded so well with small-arms and heavy ordnance should not be adapted to field artillery. But, whilst Mr. Lancaster was engaged perfecting his large shells, Sir William Armstrong was directing his energies towards improving field-guns, and the success which has attended his labours demonstrates the advantage of continually investigating each separate part of gunnery, in connexion with the science of the day; for the Armstrong gun is but the result of the studies of numberless artillerists during the last twenty years, improved

* Howard Douglas, Naval Gunnery; edit. 1851.
† Ibid.; edit. 1855. ‡ Ibid.

and put together by a practical mind with the assistance of the English mechanic. Cavalli, the Sardinian, and Wahrendorff, the Swede, brought breech-loading of rifle ordnance to great perfection; Captain Blakely, R.A., basing his inquiries on the researches of a Barlow and a Treadwell, took out a patent in 1855 for constructing guns "of an inner tube of steel, with other tubes driven or shrunk on, or with wire or bar-iron wound round the inner tube;" innumerable data exist as to the best form, number, and curve of grooves; elongated shot of cast-iron enclosed in leaden coatings have been submitted to the Ordnance Select Committee years ago; General Shrapnel, R.A., introduced the celebrated shells which bear his name, and Captain Boxer, R.A., has improved upon them; Captain Moorsom, R.N., perfected concussion fuzes, and Borman, of the Belgian service, has done much towards simplifying time fuzes. Taking, therefore, each individual part of Sir W. Armstrong's gun, there is not much that is novel in it; but the manner in which he has brought these several parts to act one with another is not only novel, but has given to field artillery a range and an accuracy far more formidable in comparison with the gun hitherto in use, than the Enfield rifle is to the English musket.* But here

* Proposed New Method of Constructing Cannon, by Captain Blakely, R.A. Pamphlet, 1858. Remarks on the Diaphragm

let us bear in mind that, if it has taken years to produce results from the field-gun proportional to those of the rifle, there is still much to be overcome before we can produce proportional effects from the siege-gun. And to show how much remains for artillerists to explore, we may add that gunpowder itself is so far from having been brought to perfection, that at the present day it is the great impediment to accuracy in rifle-shooting,—a fact forcibly exemplified by the failure of the Whitworth rifle. For some time past the daily press has informed the world that the distinguished machine manufacturer, Mr. Whitworth, has patented a military rifle, totally eclipsing the Enfield. Now any one acquainted with fire-arms knows that Mr. Whitworth has done no such thing. There is nothing new in an hexagonal bore, nor in pewter having greater power of penetration than lead; or, as pointed out many years ago, "that maximum range demanded maximum practical elongation of projectile;" or, in other words, that diminishing the diameter of the bore gives increased accuracy, as it is then easier to proportion the diameter, length, and weight of bullet to the diameter,

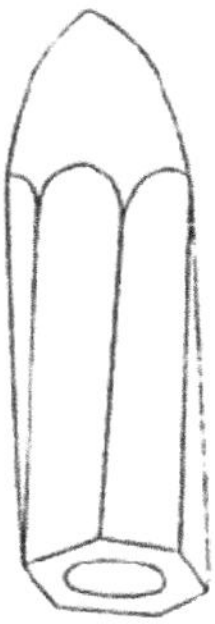

Shrapnell Shell, by Captain Boxer, R.A. 1858. Naval Gunnery, by Sir H. Douglas, Edit. 1851, p. 246. Speech of Sir W. Armstrong in "The Times," May 14, 1859.

length, and weight of barrel;* neither is there anything new in the theory, that the projectile and the inside of the barrel should be so constructed that the one should fit exactly into the other,† and thus be endowed with the most powerful rotation which can be given to it. Unfortunately an obstacle exists which renders it impossible to make this accuracy practicable—powder; for at the present day the very best, when ignited, will leave a residue, that is to say, a portion which will not consume, and this residue varies more or less according to its quality, temperature of the weather, &c. Any one curious on the subject will find a most able analysis of this matter in No. 103 of the "Philosophical Magazine" for July 1858;‡ but, as far as we are concerned, it is not necessary to enter further on the question than to say that of this residue a certain portion is blown out at the muzzle with the force of explosion, but the rest cakes round the breech and up the sides of the barrel.§

* This has been well exemplified in the half-inch bore, on the Enfield pattern, proposed by the Inspector of Small Arms at Enfield, Mr. James Gunner.

† Whitworth Patent, A.D. 1855, No. 903.

‡ On the Chemical Theory of Gunpowder, by R. Bunsen and L. Schischkoff.

§ This deposit at the breech with the present government powder averages about half an inch for every 100 rounds of musket ammunition, so that at the end of the 100 rounds you cannot ram home to within half an inch of bottom of barrel.

By the use of grease or other ingredients with the cartridge, this fouling up the barrel is got rid of to a certain extent, but in very hot weather it adheres firmly, and, after a time, prevents a man loading except with difficulty, even with ordinary windage. For this reason foreign governments have considered that our service bullet fitted too closely, and they have sacrificed still further than we have done the accuracy of shooting to the rapidity of fire, for the fire of sharp-shooters is an occasional thing in war, that of masses the constant; and when you fire at a mass it is all the same whether you hit the man you aim at or his neighbour. Now, with Mr. Whitworth's very close-fitting bullet, a soldier in action might find himself possessing a weapon down which he could not ram his bullet, and it is for this reason that the smooth bore of Lancaster's rifle presents advantages over the Enfield rifle, the grooves of which encourage this fouling. Before, therefore, we can reduce our windage we must improve our powder, and this matter will not only try the feelings of the Chancellor of the Exchequer, but many a practical manufacturer, many an ardent inventor and patentee, before it is realised. In the meantime, breech-loading appears to present decided advantages with respect to this fouling of rifled arms, for the cartridge, not having to be rammed down, the bullet can be allowed less windage ; but, like many other scientific subjects, the necessity for a thing and its execution are two very different matters; and to discover a simple mode of loading at the breech, and yet prevent the escape of the gaz of the powder when ignited, has, perhaps, occupied as many brains, and resulted in as many failures, as any subject I am aware of. For though perseverance is beginning to tell so far, that our cavalry are already armed with breech-loaders, and that amongst deserving English competitors may be mentioned Westley-Richards, Terrey, Lancaster, &c., yet, to show how much still remains to be done, we may state that a board of officers of the United States Artillery, which assembled at West Point, New York, July 12, 1858, to report on the many ingenious systems of breech-loading in that country, stated that, "Having made a careful examination and trial of the several models presented to them,* so far as this can be done without actual service in the field, the board are led to the conclusion that none of these arms are free from serious defects, which would render them unsuitable for use when subjected to the ordinary accidents and exposure of military service."

* Amongst those tried were Burnside's, Sharpe's, Starr's, Maynard's, Merrick's, Colt's, Gibb's, Symmes', Morse's, Wells', and Joslyn's.

CHAPTER VII.

Brief as has been the foregoing sketch, nevertheless I trust I have clearly demonstrated that, even with respect to its armaments, a nation cannot stand still, but must keep up with the science of the day; and that each telling improvement in this science is not the simple result of sudden thought in this or that man, but of the careful research of innumerable brains, which, as they grow old and impotent, are being ever replaced by the young and the able. And it is for these reasons, when we take into consideration the defences of this country, we must be careful not to place our sole reliance upon Government factories, which can have neither the same inducement or the same opportunity of improving on established patterns as men engaged in trade. Thus the competition of the iron-foundries, and the constant experiments which they carry on, from casting an improved rail to that of a gigantic mortar, adds a fund of experience which cannot be acquired in Woolwich Arsenal; for if one firm becomes celebrated for its cast steel and another for its malleable iron, this celebrity is the result of many a long and expensive trial, on the success of which depended the fortune of the owner. And when we think of the number of these steel and iron works, and yet how seldom a *boná-fide* improvement affects the mass of them, we may judge how little can be effected by a single Government factory, isolated by position and by interests from the iron trade. And so it is with small-arms. Look over the Patent List,* and see how many minds have been occupied in conning over the mechanical theory of the lock, the form of the bullet, the rifling of a barrel, the purifying of the iron of which the barrel is made—even the curve of the trigger or butt-plate of a musket, and it will be at once seen how necessary it is to maintain the small-arm trade in all its integrity. Take, as an illustration of this fact, the revolver pistol. There is nothing new in the idea. Pepys tells us "of a gun to discharge seven times; the best of all devices that I ever saw, and very serviceable, and not a bauble; for it is much approved of, and many thereof made;"† and the United Service Institution possesses a revolver of the time of Charles I.,

* Abridgements of the Specifications relating to Fire-arms, &c. printed by order of the Commissioners of Patents, 1859; one of a series of publications which reflects the greatest credit on the Patent Office.

† Diary, vol. ii. p. 2.

identical in principle with that of Colt's. Yet would we never have heard of the latter, had it not

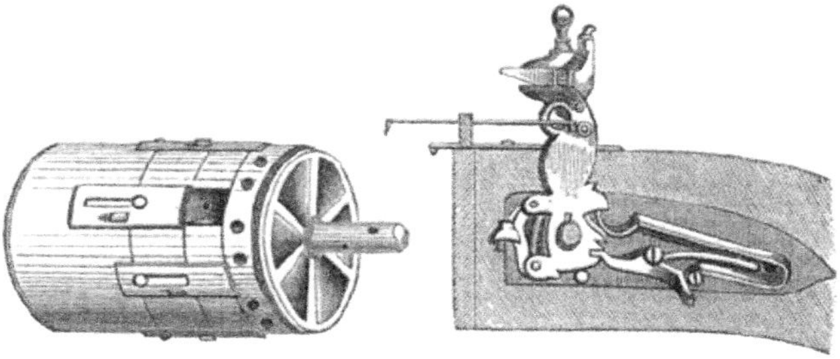

been for that marvellous emigration to the gold diggings of California and Australia, which created a demand for a powerful portable weapon of de-

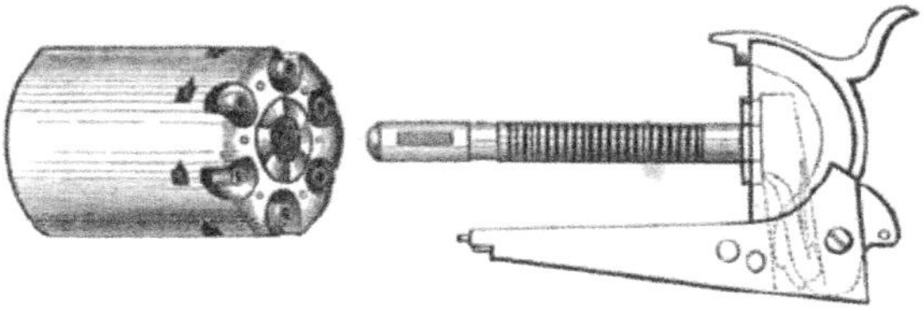

fence, and caused not simply the revival of an old idea, but also its practical solution, by the aid of modern machinery; so that a man can now carry in his waist-belt a weapon to be relied on, similar in principle to that which two centuries ago not only required a holster, but was most likely to be out of order when wanted.

That Government factories should be erected

as wholesome restrictions on the selfishness of private trade is a point which no man will dispute; but we are speaking here of an object far nobler than the doubtful saving of a few shillings—the maintenance of this empire; and the fabricating of arms* becoming identified with the commerce of a country ever affords it a powerful resource, which we of all nations can the least neglect; for, small as is the number of our standing forces and of our fortresses, we must take care they derive every assistance which art can furnish. Besides, with reference to this subject, there is another most vital point to be considered, viz. the positive danger of accumulating our establishments for manufacturing military stores within one arsenal. The dispersion of the military resources of a country over its surface, as far as possible, is one of the most important of military maxims. France has her small-arm manufactories at St. Etienne, Tulle, and Châtelherault; powder-mills at Es-

* I, of course, do not here allude to the manufacturing of laboratory stores, viz., caps, bullets, &c., the mere cost of inspecting the correctness of which would almost equal the original value of the article, but simply to iron ordnance, small arms, powder, &c., for which England has obtained a world-wide celebrity. And as a proof of the necessity of thus encouraging our trade, I may add that the sword-trade in England, on the breaking out of the Crimean war, had been nearly annihilated for want of stimulus, and is at the present moment incapable of supplying the wants of the country.

querdes, St. Ponce, Metz, Vonges, St. Chamas, Toulouse, Angoulême, St. Médard, Pont de Buis, Maromme, Le Bouchet, and Le Ripault; and military arsenals at Strasburgh, Toulouse, Douai, Auxonne, Rennes, and Metz, the three first of which possess cannon-foundries. Prussia has her manufactories of small-arms at Saarn, near Dusseldorf; at Suhl, in the mountains of Thuringia, at Potsdam and at Dantzic; and arsenals at Berlin, Posen, Neiss, Dantzic, and Deutz. Russia has exhibited a similar prudence; and if Austria has accumulated her military stores round Vienna, it is because Vienna is the centre of her empire. England alone possesses an arsenal the destruction of which would utterly annihilate for the time her resources in munition of war, and which, therefore, is to an enemy worth the loss of an army, and yet which is totally undefended by art. As to how that defence is to be effected, is a question which will no doubt induce many to advocate plans differing greatly one from the other. For, at the present day, practical fortification is far from being the defined system which is taught at military schools. We there hear a great deal of Vauban's system. Vauban had no system. Having assisted at the repairs of 300 fortresses, and having erected thirty-three new ones, we can find no two alike. In the plans of the earlier Italian engineers we find system; in those of Montalembert and Fergusson we find system; for these men created plans, and, according to these plans, they wished to fortify all places alike. Vauban's only plan was to strengthen the spot pointed out to him, so that the greatest advantage might be taken of the nature of the locality, in connexion with the weapons of the day; for the whole theory of fortification consists in the axiom, "that a body of men should be able to defend a position against a superior force until measures can be taken for its relief;" that is to say, fortification is—time. It matters not, therefore, whether the weapon of the besieger is the mangonel of the dark ages, the great gun of the middle ages, the 24-pounder of the last century, or the 68-pounder of the present, the able engineer will ever bring defence on a par with offence, and bear in mind the Todtleben system of fortification,—science, life, and money; the two last by no means the least important. For when, owing to the power of gunnery, the old city wall no longer afforded protection, and a new system had to be introduced, two items had likewise to be thought of—first, money to build them; secondly, men to defend them; and the selection, therefore, of such cities as were most suitable for this purpose became a question difficult of solution. In Holland, it was found that short distances between the fortresses, and positions that would enable a defensive army at any time to place the country under water, answered well. Louis XIV. endeavoured to imitate this example, and at a vast expense, in spite of Vauban's remonstrances; but, like all men of genius, Vauban was before his age. In 1667, he had written to Catinat:—"You are right in saying that this number of fortresses in France is an inconvenience. If an extensive war were to arise, I fear it would be seen in the first campaign,"—a prediction which he lived to see realised, for, in 1701, he again says:—"We are reduced to the defensive, for we cannot at the same time maintain many large armies on foot, and yet keep up so many fortresses; especially as they only protect the point they occupy—heavy masses without motion; whilst a powerful army can move about and impose terror in every direction. Besides, it should be sufficient to possess five or six generals, whilst it is requisite to keep an hundred in towns which armies leave behind them without besieging."* But when Frederick II. of Prussia had proved that rapidity in movements might compensate for a mass of forces, the French acknowledged the error of their "Grand Monarque," and in 1787 a committee of general officers, amongst whom were Guibert the tactician, Fourcroi the engineer, and Gribeauval the artillery-

* Paixhains, Const. Milit. p. 151. Marmont, Esprit des Instit. Milit. p. 93, unaware of these remonstrances of Vauban, lays the blame of this futile expense on him.

man, proposed to level the outworks of eleven of these strong fortresses, to withdraw the garrisons from another eleven, and to look upon the active army as the real bulwarks of the country.* And when Napoleon conquered the Continent, in spite of the fortresses on the Rhine, the Elbe, the Oder, and the Danube, it was further acknowledged by Europe that these numberless strong positions only absorbed the finances and paralysed whole armies; and when he himself committed the same mistake, in 1813 and 1814, the disastrous results which followed from such a course forced even him to admit, that "All which is human is limited—artillery, money, good officers, good generals; all these are not infinite; and if obliged to disseminate these everywhere, you will be strong nowhere."

Thanks to our insular positions, we have not much to fortify. The protection of our commercial harbours from a *coup de main*, the completing the defences of a few such points as Portsmouth, Plymouth, and Pembroke, and fortifying Woolwich Arsenal, is about all we have to do on that head to insure comparative security. But that much must be done; and as to our engines of war, let us reward successful competitors, as we have done Sir William Armstrong, and we shall find capable men taxing their energies to keep this country a-head of all improvements in implements of warfare.

* Paixhains, ditto, p. 151.

APPENDIX A.

Not only is it very improbable that Europe has derived much of its science from China, but it is more than probable that China obtained most of its knowledge from the Westward.* For instance, like the Egyptians, their Emperor is also high-priest, and they keep their feast of lanthorns. The doctrines of Pythagoras and Epicurus, in a garbled state, are those of many of their learned sects, and their religion, besides its compound of idolatry and Brahminism, has a strong mixture of Judaism. There is a tradition amongst the Chinese Jews, that a small colony of them entered China about 200 B.C.,† and the Christian religion is said to have been preached there in the seventh century,‡ statements which are strongly corroborated by many of the sayings in their "Five Books." For instance, in the book Tkin it is said that "all the oxen killed by Chew were not worth the meanest offerings of Veng Vang; because the former offered his sacrifices with a heart polluted by crimes, whereas the better part of the latter's offerings consisted in the purity of his heart."§ "It is in vain," says the Shi-king, "to hide one's self in the dark, nothing is hidden from Shang-ti, the night is as clear as day; he penetrates into the most secret corners, where the malignity of man's heart could withdraw itself from his sight; he is present everywhere,

* See a curious paper by S. De Sacy, in the "Notices et Extraits de la Bibl. du Roi," tom. xii. p. 277.

† Lettres Edifiantes, tom. vii.

‡ Du Halde, Hist. de la Chine.

§ Psalm li. ver. 15-17.

and darts his light into the most obscure windings of the most impenetrable labyrinth where man would attempt to conceal himself."* Whilst in an edict published 1700 A.D., in answer to a question which had been put by the Jesuits, the Emperor declares that they called their god Shang-ti, or Supreme Lord, because that through respect none dare call him by his proper name.†

Such remarkable coincidences between their philosophical creeds and the Scripture of the Jews attract at once the attention of any unbiassed mind; but, perhaps, is there no more convincing proof of the total inability of the Chinese to maintain that position in the history of science which some have endeavoured to place them in, than the melancholy picture drawn of it by their most ardent supporter, Abel Rémusat, in his "Discours sur l'Etat des Sciences Naturelles chez les Peuples de l'Asie Orientale, Melanges Posthumes." The fact is, that at the present day we cannot overlook such discoveries as that made by Babbage respecting their title to the discovery of logarithms, being based merely on their having translated Vlacq's tables,‡ with all its errors; or the statement of Father Mailla, that the error respecting the early use of cannon by them arose from his having been misled by their making use of nearly the same word to denote the guns of his time as the machines they made use of in the middle ages,§ a fact which is by no means confined to the Chinese; for the English word gun is only a corruption of the Μάγγανοι of the Lower Empire,‖ and musket a translation of *zimbourek* (small wasp), a species of bow made use of by the Moguls, and which Marino Sanuto first translated by the word Muschetta.¶

* Psalm cxxxix.

† Exodus, vi. 3.

‡ Wall, Orthog. of the Jews, vol. ii.

§ Letter VI. p. clxxviii.

‖ Ducange, Glossarium; Thomas Walsingham, pp. 226, 398.

¶ Quatremere, Hist. des Moguls, p. 286.

APPENDIX B.

A MANUSCRIPT preserved in the Imperial Library at Paris (No. 7239), known as a military treatise by one Paulo Santini Ducensis, gives us the most decisive information respecting the Italian artillery of this period. It was obtained by Girardin, French Ambassador at Constantinople in 1687, from amongst the works preserved at the Seraglio, and sent to Paris, when, without further inquiry, it was declared to be a manuscript of the fourteenth century, which had been written by an Italian engineer in the service of some King of Hungary, and had been taken in some expedition which the Turks had made into that kingdom. Venturi, well acquainted with Italian manuscripts, put it down as having been written about 1440; but some modern writers have supported the original suggestion, that it represented the artillery of the fourteenth century, and amongst others M. Favé, of the French Artillery, and the Emperor Louis Napoleon, have used its illustrations to that effect.* Venturi was right. The researches of Carlo Promis have shown that this Paulus Santinus was a scribe of the middle of the fifteenth century,† and the embellishments of its pages plainly show it must have been written in the second half

* Rénaud et Favé, Du Feu Grégeois, &c. Paris, 1845; Etudes sur l'Artillerie.

† Dell' Arte dell' Ingegnere, &c. Torino, 1841.

of the fifteenth century. For instance, the armour, a Cupid sitting on a dolphin, a figure of Minerva, a Venus, two large outlines of the horses of Venice, which display the pencilling peculiar to a great artist; but, above all, in fol. 71, is a copy of Pisanello's medal of Francesco Sforza, Marquis of Cremona, to celebrate his marriage, which did not take place till 1441.* I was also particularly struck with a number of little drawings which are scattered over almost every page, bearing strong characteristics of the ingenuity and humour of Da Vinci; and the more these are compared with the sketches in his manuscripts, especially Gerli's collection, the more this becomes apparent. For instance, an ostrich (fol. 66), swimming belts (fol. 82, 83), a face of an old man on the shield of a soldier ascending a scaling-ladder (fol. 102), a man cutting a chain by placing an axe on it while another strikes the axe (fol. 83), a subject to which he often reverts in his manuscripts. This induced me to search for traces of his right to left writing, and I discovered it in the chain armour of a helmet (fol. 109), and the word *salto* on the bridge in fol. 102. That this copy, therefore, once belonged to him is not improbable, and there is no necessity for its having reached Constantinople through Hungary. Sigismund Pandolpho † sent a copy of Valturius' work, "De Re Militari," illustrated by Matteo Pasti of Verona, to Mahomet II.,

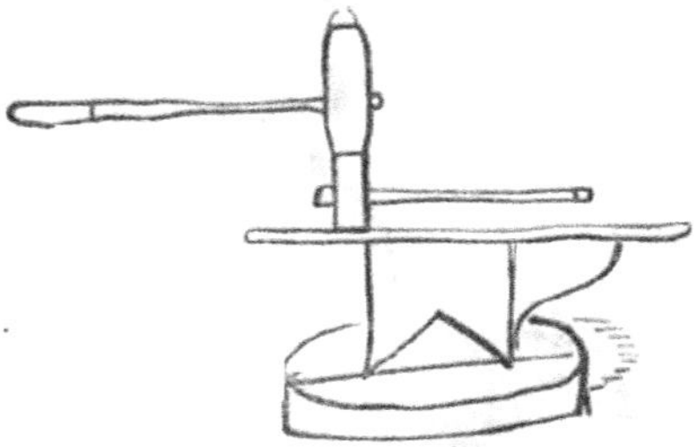

Leonardo da Vinci, G. fol. 7.

* A *facsimile* of this medal is to be found in Litta's Italian Genealogies, Sforza family.

† His letter may be found in the Miscellenea of Stephanus Baluzius.

who, for a Turk, was particularly fond of European art; and there is no reason whatsoever for supposing that this manuscript did not reach Constantinople in as legitimate a way.

Its publication would be of interest, not only to the military antiquarian, but as exemplifying the simple state of engineering of this period.

APPENDIX C.

THESE note-books are principally small pocket memorandum books; twelve of them, marked A, B, &c., are preserved at the Institut de France. A thirteenth, N, is in the Ambrosian Library at Milan. They are written from right to left, which has induced many to suppose he was left-handed: but in I, fol. 59, G, fol. 88, H, 3d part, fol. 46, are specimens of a fine bold writing from left to right.* In any one of them may be found heaped up together political apologues, philosophical maxims, scientific researches on the anatomical nature of the flight of birds, questions on botany, mechanics, dynamics, hydraulics, a sonnet, researches on languages. Indeed, I cannot better exemplify their confusion than by stating that in the midst of dynamics I found "The dutchess' bath should be three-quarters hot, one-quarter cold" (I, fol. 34). The publication of these manuscripts just as they are, which could easily be done by means of photography, together with a French translation, would be well worthy of the French government, and would throw a good deal of light on the state of science in the beginning of the sixteenth century. It is said that M. Libri intended to bring out an entire edition of Leonardo da Vinci's works, but on perusing that author's "Histoire des Sciences Mathé-

* This does not seem to have been perceived by Lomazzo or Vasari.

matiques en Italie," I felt much annoyed, as many others must have done, at the very imperfect manner in which he has carried out that portion of his work relative to Da Vinci. As chief of the Great Library in the Rue Richelieu and a member of the Institut, M. Libri had ample opportunity of giving to the world the results of Leonardo's mathematical studies, but the difficulty seems to have dismayed him; for, after raising the curiosity of the reader to the highest pitch, he merely places before him a few notes copied letter for letter from the original, and therefore perfectly unintelligible to most people. Even though his intention was, as he says, to give to the world a separate work on Leonardo, still he might either have translated that part he did publish into the same French as the remainder of the volume, or put it into modern Italian. Besides, he gives little beyond what Venturi* had already done, for his own studies of the note-books seem to have been very superficial. For instance, he brings him forward as the discoverer of the algebraic signs + and —. The slightest attention would have shown him that the + he saw was only a very careless 4, and the — a fractional mark of division.

* Essai sur les Ouvrages Physico Mathém de Leonardo da Vinci.

APPENDIX D.

A VERY good account of English ordnance of the eighteenth century will be found in "Rees' Cyclopædia," A.D. 1819, Art. Cannon. But it must be borne in mind that copper was not discovered in England till towards the end of the seventeenth century, when, in 1691 A.D., the English Copper Company was formed, and in 1702 the first brass-work was erected at Bristol. Previous, and indeed up, to 1750, England was chiefly dependent on Germany and Sweden. In 1694 the English Government paid as high as 1*s.* 6*d.* per lb. for Swedish copper, and in 1717 English copper was 147*l.* per ton. From 1720–72 cake copper was 130*l.* to 100*l.* per ton. The discovery of fresh mines in Derbyshire and Wales caused copper to fall in 1791 to 79*l.* per ton. France was chiefly supplied from the Armenian, Swedish, and South American mines. (Parl. Rep. on the State of the Copper Mines and Copper Trade, 1779 A.D.)

London:—Printed by G. BARCLAY, Castle St. Leicester Sq.

Related Titles from Westphalia Press

The Limits of Moderation: Jimmy Carter and the Ironies of American Liberalism
by Leo P. Ribuffo

The Limits of Moderation: Jimmy Carter and the Ironies of American Liberalism is not a finished product. Yet, this book is a close and careful history of a short yet transformative period in American political history, when big changes were afoot. and continue to shape our world.

Bunker Diplomacy: An Arab-American in the U.S. Foreign Service
by Nabeel Khoury

After twenty-five years in the Foreign Service, Dr. Nabeel A. Khoury retired from the U.S. Department of State in 2013 with the rank of Minister Counselor. In his last overseas posting, Khoury served as deputy chief of mission at the U.S. embassy in Yemen (2004-2007).

Energy Law and Policy in a Climate-Constrained World
by Victor Byers Flatt, Alfonso López de la Osa Escribano, Aubin Nzaou-Kongo

This book presents reflections on concepts, foreign policy, regional and international cooperation, and the specific role the state is to play when it comes to such thing as energy law and policy.

The Zelensky Method
by Grant Farred

Locating Russian's war within a global context, The Zelensky Method is unsparing in its critique of those nations, who have refused to condemn Russia's invasion and are doing everything they can to prevent economic sanctions from being imposed on the Kremlin.

The Lord of the Desert: A Study of the Papers of the British Officer John B. Glubb in Jordan and Iraq
by Dr. Sa'ad Abudayeh

John Bajot Glubb, a British engineer officer, was sent to Iraq in 1920 to resolve the problems which erupted after the Iraqi revolt. He remained in the area for ten years, working with the Bedouins. In 1930, he moved to Jordan for twenty-six successful years. He invented what Dr. Abudayeh calls the Diplomacy of Desert.

Managing Challenges for the Flint Water Crisis
Edited by Toyna E. Thornton, Andrew D. Williams, Katherine M. Simon, Jennifer F. Sklarew

This edited volume examines several public management and intergovernmental failures, with particular attention on social, political, and financial impacts. Understanding disaster meaning, even causality, is essential to the problem-solving process.

Resistance: Reflections on Survival, Hope and Love
Poetry by William Morris, Photography by Jackie Malden

Resistance is a book of poems with photographs or a book of photographs with poems depending on your perspective. The book is comprised of three sections titled respectively: On Survival, On Hope, and On Love.

Abortion and Informed Common Sense
by Max J. Skidmore

The controversy over a woman's "right to choose," as opposed to the numerous "rights" that abortion opponents decide should be assumed to exist for "unborn children," has always struck me as incomplete. Two missing elements of the argument seems obvious, yet they remain almost completely overlooked.

Geopolitics of Outer Space: Global Security and Development
by Ilayda Aydin

A desire for increased security and rapid development is driving nation-states to engage in an intensifying competition for the unique assets of space. This book analyses the Chinese-American space discourse from the lenses of international relations theory, history and political psychology to explore these questions.

Made in the USA
Monee, IL
04 November 2026

67383622R00075